8/14

BARRON'S
NEW JERSEY

ASK 6
MATH
TEST

D1543435

SECOND EDITION

Mary Serpico, M.A.

About the Author

Mary Serpico has been in the field of mathematics education for more than thirty years. During her career, she has been introduced to many local, state, national, and global curriculum initiatives, and she has worked collaboratively to design, develop, and implement them in various school settings.

All inquiries should be addressed to:
Barron's Educational Series, Inc.
250 Wireless Boulevard
Hauppauge, New York 11788
www.barronseduc.com

ISBN: 978-1-4380-0050-3
ISSN: 2169–5091

Date of manufacture: December 2012
Manufactured by: B11R11, Robbinsville, N.J.

Printed in the United States of America
9 8 7 6 5 4 3 2 1

10%
POST-CONSUMER
WASTE
Paper contains a minimum
of 10% post-consumer
waste (PCW). Paper used
in this book was derived
from certified, sustainable
forestlands.

CONTENTS

CHAPTER 4: STATISTICS AND PROBABILITY / 121

CHAPTER 5: RATIOS AND PROPORTIONAL RELATIONSHIPS / 141

CHAPTER 6: SAMPLE TESTS / 147

APPENDICES / 199

INTRODUCTION TO THE NEW JERSEY ASK 6 MATH TEST

This spring you will be taking the New Jersey ASK 6—Assessment of Skills and Knowledge in Mathematics. This is a test that measures how well you understand and can apply all of the mathematics that you've learned so far. The test is based on the Common Core Standards and measures your knowledge of them. The domains include:

> The Number System
> Geometry
> Expressions and Equations
> Statistics and Probability
> Ratios and Proportional Relationships

The NJ ASK 6 Math Test is given over two days. You will be asked to answer short constructed-response questions, multiple-choice questions, and extended constructed-response or open-ended questions. There will be a combination of questions taken from the topics listed above. Some questions will require that you use two or more topics and solve problems that have multiple parts. You may be asked to explain your reasoning and processes. The tests will last for about one hour each day.

The New Jersey ASK 6 tests both computation and your ability to solve mathematics problems. You may use a calculator for some sections of the test. This will allow you more time to focus on problem solving and application. In addition, you will be given a reference sheet of the important formulas and information, so it is not necessary for you to memorize them. A sample reference sheet is included at the end of this book.

In some cases, other materials such as a ruler, protractor, or shapes may be provided if they are needed for any part of the test. Your teacher will explain how to manage these tools as you go over the directions. A sample of these materials are provided for you also.

As you prepare for the New Jersey ASK 6, remember to use the following test-taking skills:

- Listen carefully and follow all directions.
- Read each question carefully, to make sure that you know what you are being asked to do.
- Fill in and darken the spaces for the multiple-choice questions completely and make sure that they are in the correct spaces on the answer sheet.
- Use your reasoning and estimation skills to eliminate incorrect answers that cannot be correct.
- Skip any questions that you do not know and go on to the next question. You can go back and check a section if there is time.
- Think about what you are being asked to do in the open-ended questions. You are being scored on your ability to explain mathematics, not your ability to write a paragraph. Use appropriate mathematical symbols and expressions.
- Arrive for the test after a good night's rest and a nutritious breakfast.
- Be confident that you will do your best!

THE NUMBER SYSTEM

The number system relates to all types of numbers. In the sixth grade, we work with whole numbers, fractions, decimals, common percents, and both positive and negative integers.

LESSON 1—WHOLE NUMBER PLACE VALUE AND COMPUTATION

The set of whole numbers starts with zero and never ends. Each of the digits in a whole number is named by a certain place value.

The following place value table can be helpful when you are recognizing, reading, and writing whole numbers.

PLACE VALUE TABLE FOR WHOLE NUMBERS

Hundred Millions	Ten Millions	Millions	Hundred Thousands	Ten Thousands	Thousands	Hundreds	Tens	Ones
		5	4	3	9	0	2	6

In the place value table, the number five million, four hundred thirty-nine thousand, twenty-six (5,439,026 written in standard form) has each of its digits in the corresponding column. The value of the nine in the thousands column is 9,000, and the value of the four is 400,000 because it is in the hundred-thousands column.

Notice that there are no hundreds in this number, which is why a zero is used as a placeholder in the hundreds column.

Try This

Multiple-Choice Questions

Answer the following using the number 1,489,653. Use the place value table if necessary.

1. What is the value of the 6?

    A. 6

    B. 60

    C. 600

    D. 6,000

2. What is the value of the 1, written in standard form?

    A. 1,000,000

    B. 10,000,000

    C. 1 million

    D. one

3. What is the value of the digit in the ten-thousands place?

    A. 8

    B. 8,000

    C. 80,000

    D. 800,000

4. How many hundred thousands are there?

    A. 1

    B. 2

    C. 3

    D. 4

Short Constructed-Response Question (SCR)

What is the largest nine-digit whole number that has a one in the thousands place and a six in the tens place? You can only use each number once.

The answers appear on page 201.

COMPARING AND ORDERING WHOLE NUMBERS

We use place value to help us compare whole numbers to determine whether they are greater than, less than, or equal to each other. We use the symbols <, >, and =. To order numbers means to list them from least to greatest or from greatest to least.

Think About This

Express the following sets of numbers in two different ways.

1. 6,421 and 6,412

2. 12,589 and 32,654

3. 15,699 and 15,799

Order each set of numbers from least to greatest and greatest to least.

4. 76, 85, 29, 15, 63, 46

5. 290, 920, 209, 902, 29

6. 1,598, 2,365, 1,589, 1,745, 2,662

The answers appear on pages 201–202.

ROUNDING WHOLE NUMBERS

It is important to reason with numbers by predicting a reasonable answer and checking the reasonableness of it once you solve the problem. Let's look at some different strategies to do this.

Rounding numbers allows us to work with numbers that are easy to understand. For example, if a video game costs $68.47, it is more reasonable to think of it as about $70.00. Another example is that if you are bringing refreshments for eighteen students in your class, you might round that number to twenty.

A good way to round whole numbers is to determine which two consecutive multiples the number is closest to. For example, 368 is between 300 and 400. Because 368 is closer to 400, then 368 rounded to the nearest hundred is 400. When rounding to the nearest ten, 368 is between 360 and 370, so it rounds to 370.

Consecutive multiples are found by skip counting. The multiples of ten are 10, 20, 30, 40, and so on. Consecutive multiples are the multiples that follow in order. For example, 10 and 20 are consecutive, 20 and 30 are consecutive, 30 and 40 are consecutive, and so on. When you round, you decide which two consecutive multiples the number is closest to.

Think About This

1. The first two consecutive multiples of 10 are 10 and 20. Name the next three.

2. What are the first five consecutive multiples of 100? 1,000?

3. Tell which two consecutive multiples of one hundred these numbers come between and round each of them to the nearest hundred.   135   643

4. Tell which two consecutive multiples of one thousand these numbers come between and round them to the nearest thousand.   3,235   6,941

5. Tell which two consecutive multiples of ten thousand these numbers come between and round to the nearest ten thousand.   31,786   76,234

The answers appear on page 202.

ROUNDING DECIMALS

When we round decimals, we follow the same process as we use for fractions. If the decimal value is less than one, we round it to zero, five-tenths (which is equivalent to one-half), or one. Mixed decimals are usually rounded to the nearest whole number.

Let's look at some examples. The decimal 0.85, or eighty-five hundredths, is between five-tenths and one. Since it is closer to one, we round it to one. The decimal 0.21 is between zero and five-tenths. Since it is closer to zero, we round it to zero.

Think About This

Are the following decimals closer to zero, one-half, or one?

1. 0.123

2. 0.876

3. 0.6432

Round these mixed decimals to the nearest whole number:

4. 3.987

5. 5.0456

6. 135.876

The answers appear on page 202.

Decimals as Money

Money is the most important application for decimals. Our monetary system is based on mixed decimals through hundredths. Dollars are represented by the whole numbers, and the decimal part of the expression represents the change or cents. It is very important to be able to estimate the reasonableness of the amounts of money that we spend or earn. We use all of our rules for rounding when we are working with money. If the values are small, we apply the methods that we use with fractions and decimals. If we are working with large amounts of money, we apply the processes we use with whole numbers.

Let's look at some examples.

Think About This

1. Erasers each cost $0.35 in the school store. Does that round to zero, fifty cents, or one dollar?

2. Basketball game tickets each cost $3.75. Does that round to $3.00 or $4.00?

3. You raised $87.00 for the school fundraiser. Does that round to $80.00 or $90.00?

4. The cafeteria sold $1,875.00 worth of lunches. Does that round to $1,000.00 or $2,000.00?

5. Your sister's college tuition is $28,825.50. Does that round to $20,000 or $30,000?

6. Your parents have a $120,000.00 mortgage. Does that round to $100,000 or $200,000?

7. Give two examples of amounts of money for each that round to zero, $1.00, $10.00, $100.00, $1,000.00, $10,000.00, $100,000.00, and $1,000,000.00.

The answers appear on pages 202–203.

Estimation in Computation

We round numbers to estimate what our answer should be before we do any computation, or to verify the reasonableness of our answer once we have calculated it.

For Example

There are five sixth-grade classes in a team. The enrollment in each class is 23, 24, 25, 26, and 27. We want to know the total enrollment. Before we enter any numbers into a calculator, we are going to estimate our answer by rounding and adding $20 + 20 + 30 + 30 + 30$ and getting a sum of 130. We can now expect that our answer is about 130. When we solve for the exact answer, it is 125. This is close to our estimate, and is therefore a reasonable response.

Think About This

You want to download 18 songs onto your MP3 player and each song costs $1.09. How will you decide if you are being charged the right amount?

The answer appears on page 203.

Try This

Multiple-Choice Questions

1. Which two consecutive multiples does the number 485 come between?

   A. zero and 500

   B. 400 and 500

   C. 450 and 460

   D. 1,000 and 2,000

2. Round 23,765 to the nearest hundred.

   A. 20,000

   B. 23,700

   C. 23,800

   D. 24,000

3. Which set of numbers can always be rounded to zero, one-half, or one?

   A. whole numbers

   B. mixed fractions

   C. integers

   D. proper fractions

4. What is the best rounded value for $3\frac{3}{8}$?

   A. 0

   B. 3

   C. $3\frac{1}{2}$

   D. 4

5. What is the best rounded value for 27.125?

   A. 27

   B. 27.1

   C. 27.5

   D. 28

6. What is the best estimate for 21 tickets that cost $4.75 each?

    A. $80.00

    B. $99.75

    C. $100.00

    D. $200.00

Open-Ended Questions

Fill in the missing information and explain the process of determining the reasonableness of an amount.

The youth group wants to raise $1,075.00 so that everyone can go to Great Adventure. They decide to sponsor a car wash and charge $5.25 per car. The youth group advisor estimates that they should wash about 400 cars so that everyone can go. Complete the following process to determine the reasonableness of that estimate.

1. Round to the nearest hundred the amount of money that they want to raise.   _____

2. Round the cost of washing each car.   _____

3. What operation will you use to determine how many cars they must wash?   _____

4. Estimate the number of cars.   _____

5. Calculate the exact number of cars. Round to the nearest hundredth.   _____

6. Compare the results and write a letter to the youth group advisor explaining why you agree or disagree with the estimate.

The answers appear on page 203.

NUMERICAL OPERATIONS WITH WHOLE NUMBERS

It is important that you know how to add, subtract, multiply, and divide all types of numbers, and when to use each operation to solve problems. Let's look at some examples to understand what this means.

■ Four students brought in permission slips for the field trip on Monday, and seven students brought their permission slips in on Tuesday. How many students brought their permission slips in altogether?
Use *addition* to solve.

■ The fifth grade collected 625 pull-tabs for the school fundraiser, while the sixth grade collected 1,225. How many more pull tabs did the sixth grade collect?
Use *subtraction* to solve.

■ The auditorium has 35 seats in each row, and there are 47 rows in the auditorium. How many seats are there altogether?
Use *multiplication* to solve.

■ Pat bought eight CDs for a total of $120.00. How much did each CD cost?
Use *division* to solve.

Before you can begin to solve a problem, you have to *decide which operation to use*. Then you must operate on the numbers correctly to solve for the sum, difference, product, or quotient.

Think About This

Which operation would you use to solve?

1. Thomas worked 32 hours last week and earned $7.25 per hour. How much did Thomas earn?

Operation: _____

Solution: _____

2. Four hundred sixteen students played on twenty-six different basketball teams. How many students were on each team?

Operation: _____   Solution: _____

3. Jessica is inviting 16 people to her birthday party. She mailed 11 invitations so far. How many more invitations does she have to mail?

Operation: _____   Solution: _____

The answers appear on page 204.

WHOLE NUMBER COMPUTATION

You should have mastered the process of adding and subtracting multidigit whole numbers by now.

Try some multiplication practice with these examples.

Try This

1. 345 ×28	**2.** 6,274 × 197	**3.** 76,348 × 806
4. 7,098 × 645	**5.** 2,111 × 987	**6.** 36,402 × 312

Now, let's move on to division. Divide 968 by 33.

$$\begin{array}{r} 29 \\ 33\overline{)968} \\ \underline{66} \\ 308 \\ \underline{297} \\ 11 \end{array}$$

(Think of the consecutive multiples of 33.)

Estimate, and divide how many times 33 can go into 96.

Multiply and subtract.

Subtract and compare (make sure the difference is less than the divisor).

Bring down the number in the ones column. Repeat the process by thinking of the consecutive multiples of 33 and estimating how many times 33 can go into 308. Multiply, subtract, and compare.

Notice that after we completed the division process, there is a remainder of 11. We can represent this quotient in different ways. We can say that $968 \div 33 = 29$ with a whole number remainder of 11, or that $968 \div 33 = 29$ with a fraction remainder of $\dfrac{11}{33}$ or $\dfrac{1}{3}$, or that $968 \div 33 = 29.33$, with a repeating decimal remainder.

Try This

Divide and represent the quotients with remainders as whole numbers, fractions, or decimals.

1. $885 \div 25$

2. $306 \div 72$

3. $28,215 \div 95$

4. $25\overline{)40,000}$

5. $123\overline{)11,193}$

6. $26\overline{)29,809}$

The answers appear on page 204.

SQUARES AND CUBES

Another way of representing multiplication is by using exponential form.

Let's look at this example:

The expression 6^2 is read "6 to the second power," or "six to the power of two," or "six squared." This means that the number 6 is taken as a factor two times, or multiplied by itself, or 6 × 6.

In the expression 6^2, the number 6 is defined as the **base**, and the number 2 is defined as the **exponent**. To simplify this expression, we first write it in its expanded form, 6 × 6. Next we solve by multiplying 6 × 6, and represent the answer in its standard form, 36.

When a number is written in exponential form having an exponent of 3, such as 4^3, it is read, "4 to the third power or 4 cubed." To solve, we first represent it in its expanded form, 4 × 4 × 4, and multiply. The product standard form is 64.

Think About This

What is the expanded form and product for each expression?

1. 2^3

2. 15^2

Name the base in each of the above expressions.

Name the exponent in each of the above expressions.

The answers appear on page 205.

ORDER OF OPERATIONS

There is a system in mathematics called the order of operations that outlines the order in which we should perform the operations. The following outline tells us how to apply the order of operations.

Step 1—Do what is in the parentheses.

Step 2—Find any squares or cubes and express them in standard form.

Step 3—Begin at the left and solve any multiplication or division.

Step 4—Return to the beginning and solve any addition or subtraction in order from left to right.

Look at these examples:

Example 1

Find the value of	$3 + 4 \times (2 - 1)$
First solve inside the parentheses— everything else stays the same	$3 + 4 \times 1$
There are no squares or cubes, so multiplication comes next (4×1)	$3 + 4$
Now add the numbers that are left	7

Example 2

Find the value of	$12 + 4 \div 2 \times 3$
There are no parentheses, so look for multiplication or division	$12 + 2 \times 3$
(Remember to do them in order from left to right)—now multiply	$12 + 6$
Do the addition last	18

Example 3

Find the value of	$5 + 2 \times 3 - 21 \div 3$
Start with multiplication	$5 + 6 - 21 \div 3$
Next divide	$5 + 6 - 7$
Then add and subtract from left to right	$11 - 7$
	4

Think About This

In what order should you operate? Rewrite each step and solve.

1. $10 - 2 \times 6 \div 4 + 5$

2. $6 \times (2 + 10) \div 3$

Insert parentheses to make each answer correct.

3. $3 \times 2 + 4 \div 3 = 6$

4. $4 + 1 + 2 \times 5 = 15$

5. $4 + 1 + 2 \times 5 = 19$

6. José solved the following example $1 + 3 \times 4 - 4 \div 3$ and got 4 as his answer. When he entered the *same problem* into his calculator he got $11.66 \therefore$ . . as the answer. Which is the correct answer? Explain your reasoning.

The answers appear on page 205.

Try This

Multiple-Choice Questions

1. Which operation is used to total the number of players on each team?

    A. addition

    B. subtraction

    C. multiplication

    D. division

2. In which operation are factors and products used?

    A. addition

    B. subtraction

    C. multiplication

    D. division

3. Which is *not* a correct quotient for the problem $12 \div 5$?

    A. 2.4

    B. 2 r2

    C. $2\frac{2}{5}$

    D. 2 r4

4. In the expression 6^3, what is the name for the 3?

    A. base

    B. exponent

    C. standard form

    D. cube

5. Which is *not* a way to name the expression 4×4?

    A. 4 squared

    B. 4^2

    C. 4 to the second power

    D. 8

6. Which is the correct solution for $3 + 2 \times 8 - 3^2$?

    A. 9

    B. 10

    C. 31

    D. 37

7. Which would you do first to solve $3 \times 4 + (9 - 2)^3$?

    A. multiply 3×4

    B. subtract $9 - 2$

    C. raise 7 to the third power

    D. add $12 + 7$

Extended Constructed-Response

Ellen was asked to solve for the quotient of $328 \div 16$ with a whole number remainder. She used a calculator and wrote the answer 20 r5. Her teacher marked it wrong. What is the correct answer? Explain what Ellen did wrong.

The answers appear on page 206.

LESSON 2—FRACTION PLACE VALUE AND COMPUTATION

Another set of numbers that we work with is fractions. Proper fractions are numbers that are greater than zero, but less than one. Each fraction represents a *part of a whole*, a *subset of a set*, a *location on a number line*, or a *division of whole numbers* problem.

Let's look at some of these examples:

Part of a Whole

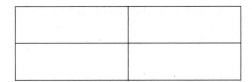

Study the rectangle. It is a whole object that has been broken into four equal parts. Each part represents one-fourth of the rectangle. We write the fraction one-fourth as $\frac{1}{4}$, naming the **denominator** as the number of parts that the object is divided into, and the **numerator** as the number of parts that we are defining. If we shade one of the parts of our rectangle, we would name it with the fraction one-fourth, and we would write it as $\frac{1}{4}$. If we shade in three of the four parts, we would name that fraction $\frac{3}{4}$.

Subset of a Set

A group of objects is a **set**, and a part of that group is a **subset** or a **fraction of a set**. For example, the math class is a set of 24 students having 10 boys and 14 girls. The fraction $\frac{10}{24}$ represents the subset of boys, and the fraction $\frac{14}{24}$ represents the subset of girls. We can simplify these fractions and say that $\frac{5}{12}$ of the class is boys, and $\frac{7}{12}$ of the class is girls.

Fractions on a Number Line

When we represent proper fractions on a number line, they will all be located between zero and one because all fractions represent part of a number.

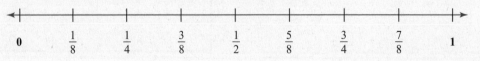

In this example, each stroke on the number line represents one-eighth of the distance between zero and one. Notice that values of the fractions are given in terms that are equivalent to eighths. For example, the distance between

$\frac{1}{8}$ and $\frac{3}{8}$ is $\frac{2}{8}$. However, it is written as $\frac{1}{4}$ on our number line because $\frac{1}{4}$ is equivalent to the fraction $\frac{2}{8}$, so we write it in simplest form.

Fractions as Division of Whole Numbers

Fractions represent division problems where the denominator is the divisor and the numerator is the dividend. When we recognize the fraction $\frac{1}{2}$ as division of whole numbers, what we really mean is $1 \div 2$. The fraction $\frac{3}{8}$ as a division problem is $3 \div 8$, and $\frac{4}{7}$ is the same as $4 \div 7$.

When we divide the numerator of a proper fraction by its denominator, the result is between zero and one, or a decimal. For example, the fraction $\frac{1}{2}$, or $1 \div 2$, is equal to 0.5. The fraction $\frac{2}{5}$, or $2 \div 5$, is equal to 0.4.

Rounding Fractions

Fractions can also be rounded to estimate answers. When we work with fractions, we usually round to the nearest whole number or one-half.

Let's look at some examples. Think about the fraction $\frac{1}{8}$.

Ask yourself, "Is it closest to 0, $\frac{1}{2}$, or 1?" Think about where it is on the number line. Since it is closer to zero, the fraction $\frac{1}{8}$ rounds to zero.

We usually round mixed fractions to the nearest whole number. Think about the mixed fraction $3\frac{3}{4}$. It is between the whole numbers 3 and 4. Since it is closer to 4, we would round it to 4.

Think About This

Round each of these fractions to 0, $\frac{1}{2}$, or 1.

1. $\frac{1}{8}$

2. $\frac{3}{8}$

3. $\frac{7}{8}$

4. Name three fractions that you would round to zero, one-half, and one.

Round each of these mixed fractions to the nearest whole number.

5. $6\frac{1}{4}$

6. $1\frac{3}{4}$

7. $7\frac{1}{8}$

The answers appear on page 206.

Try This

Multiple-Choice Questions

1. In the fraction $\frac{4}{5}$, the number four 4 is the

    A. fraction

    B. denominator

    C. numerator

    D. whole number

2. Which names the number of parts that a set is divided into?

    A. fraction

    B. denominator

    C. numerator

    D. whole number

3. There are 60 colored markers in the art box. Fifteen (15) of them are red. What is the subset of red markers in the set of markers?

    A. $\frac{1}{4}$

    B. $\frac{1}{2}$

    C. 15

    D. 60

4. Between which two fractions will $\frac{2}{5}$ be on a number line?

 A. 0 and $\frac{1}{5}$

 B. $\frac{1}{2}$ and $\frac{3}{4}$

 C. $\frac{3}{10}$ and $\frac{1}{2}$

 D. $\frac{3}{4}$ and 1

5. Which fraction names the division problem $5 \div 6$?

 A. $\frac{5}{6}$

 B. $\frac{1}{5}$

 C. $\frac{6}{5}$

 D. $\frac{1}{6}$

6. Which division problem is equivalent to the fraction $\frac{3}{12}$?

 A. $1 \div 3$

 B. $1 \div 12$

 C. $3 \div 12$

 D. $12 \div 3$

7. Which decimal is equivalent to the fraction $\frac{2}{8}$?

    A. 0.25

    B. 0.28

    C. 0.82

    D. 2.8

Extended-Constructed Response Question

Draw a number line that is divided into tenths. Label the fraction and decimal values.

The answers appear on page 207.

NUMERICAL OPERATIONS WITH FRACTIONS

Before we begin to add, subtract, multiply, and divide, let's review two sets of numbers that are related to fractions—factors and multiples.

Factors

Factors are numbers that we multiply together that result in a certain number called the **product**. In the example $3 \times 2 = 6$, 3 and 2 are the factors. 6 is the product.

Name the factors in these multiplication sentences:

1. $4 \times 8 = 32$

2. $2 \times 0 = 0$

When we work with larger numbers, we see that they have many factors. For example, the product 24 can be written as 1×24, 2×12, 3×8, or 4×6. This shows us that the factors of 24 are 1, 2, 3, 4, 6, 8, 12, and 24.

Think About This

List all of the factors of:

3. 61

4. 42

5. 125

The answers appear on page 208.

Common Factors and Greatest Common Factor (GCF)

Common factors are factors that are found in two or more numbers. To determine the common factors of two numbers, list the factors of each to see which are the same, or common.

For Example

What are the common factors of 15 and 21?
List the factors of 15 (1, 3, 5, and 15).
List the factors of 21 (1, 3, 7, and 21).

Notice that the factors 1 and 3 appear in both lists. Therefore, 1 and 3 are common factors of 15 and 21. The greatest common factor (GCF) is the largest factor that is common to both numbers. In this example, the GCF of 15 and 21 is 3.

Think About This

List the factors of each number in each set, and name the GCF.

1. 16 and 36

2. 28 and 84

3. 65 and 100

The answers appear on page 208.

Multiples

Multiple is another name for product—the answer in a multiplication problem. When we skip count, the answers are multiples. When we count by threes, the results are multiples of three: 3, 6, 9, 12, 15, 18... The multiples of 20 are 20, 40, 60, 80 . . .

Think About This

Name the first five multiples of:

1. 7

2. 10

3. 13

The answers appear on page 208.

Common Multiples and Least Common Multiple (LCM)

We find **common multiples** of numbers by listing their multiples and determining which are the same or common. For example, twelve is a common multiple of 3 and 4.

Some multiples of 3 are 3, 6, 9, 12, 15, 18, 21, and 24.
Some multiples of 4 are 4, 8, 12, 16, 20, and 24.

Both 12 and 24 are in both lists, making them common multiples of 3 and 4. In most cases, we look for the least (smallest) common multiple (LCM) of two numbers. In our example, the least common multiple of 3 and 4 is 12.

Think About This

Name the least common multiple (LCM) of:

1. 5 and 9

2. 6 and 8

3. 9 and 12

The answers appear on page 209.

Try This

Multiple-Choice Questions

1. Which is *not* a factor of 12?

   A. 1

   B. 3

   C. 4

   D. 24

2. Which factors are common to 18 and 27?

   A. 2, 3, and 6

   B. 3, 6, and 9

   C. 1, 3, and 9

   D. 9, 18, and 27

3. Which pair of numbers has a GCF of 5?

   A. 10 and 25

   B. 25 and 50

   C. 3 and 15

   D. 1 and 5

4. Which number is a multiple of 8?

   A. 1

   B. 2

   C. 4

   D. 8

5. What is the least common multiple (LCM) of 9 and 24?

   A. 1

   B. 3

   C. 5

   D. 72

6. What is the greatest common factor of 16 and 88?

   A. 1

   B. 2

   C. 8

   D. 16

The answers appear on page 209.

ADDING FRACTIONS

Now that we have recalled the foundations of fractions, we can review how to add fractions.

Remember that in order to add fractions, they must have a common denominator. It should be the least common multiple (LCM) of both addends.

Study This Example

$$\text{Add: } \frac{4}{9} + \frac{3}{4}$$

First, the LCM of 9 and 4 is 36. So the common denominator is 36.

Determine equivalent fractions for each of the addends when the denominator is 36.

$$\frac{4}{9} = \frac{16}{36} \text{ and } \frac{3}{4} = \frac{27}{36}$$

Add the numerators: $16 + 27 = 43$

The denominator is the LCM: $\frac{43}{36}$

The sum of $\frac{4}{9} + \frac{3}{4} = \frac{43}{36}$.

Simplify the answer by converting it from an improper fraction to a mixed number.

The final answer is $\frac{43}{36} = 1\frac{7}{36}$.

Try This

Add these fractions, and simplify your answers.

1. $\frac{5}{8} + \frac{3}{16}$

2. $\frac{3}{7} + \frac{4}{5}$

3. $\frac{1}{6} + \frac{8}{9}$

4. $\frac{7}{12} + \frac{4}{15}$

5. $\frac{3}{10} + \frac{7}{8}$

6. $\frac{2}{3} + \frac{11}{16}$

The answers appear on page 209.

SUBTRACTING FRACTIONS

Now let's move on to subtraction.

Remember that in order to subtract fractions, they must have a common denominator. It should be the least common multiple (LCM) of both terms.

Study This Example

Subtract: $\dfrac{8}{9} - \dfrac{3}{4}$

First, the LCM of 9 and 4 is 36. So the common denominator is 36.

Determine equivalent fractions when the denominator is 36.

$\dfrac{8}{9} = \dfrac{32}{36}$ and $\dfrac{3}{4} = \dfrac{27}{36}$

Subtract the numerators: $32 - 27 = 5$

The denominator is the LCM: $\dfrac{5}{36}$

The difference of $\dfrac{8}{9} - \dfrac{3}{4} = \dfrac{5}{36}$.

It is not necessary to simplify the answer because it is already in lowest terms.

Try This

Subtract these fractions, and simplify your answers.

1. $\dfrac{5}{8} - \dfrac{3}{16}$

2. $\dfrac{4}{5} - \dfrac{3}{7}$

3. $\dfrac{8}{9} - \dfrac{1}{6}$

4. $\dfrac{7}{12} - \dfrac{4}{15}$

5. $\dfrac{7}{8} - \dfrac{3}{10}$

6. $\dfrac{11}{16} - \dfrac{2}{3}$

The answers appear on page 210.

ADDITION AND SUBTRACTION WITH MIXED FRACTIONS

Mixed fractions are made up of whole numbers and fractions together. $9\dfrac{1}{8}$ and $6\dfrac{2}{5}$ are both mixed fractions. They are sometimes called mixed numbers.

When we add mixed numbers, combine all terms and simplify the sum.

Study This Example

Add: $2\dfrac{8}{9} + 3\dfrac{3}{4}$

First, the LCM of 9 and 4 is 36. So the common denominator is 36.

Determine equivalent fractions when the denominator is 36.

$\dfrac{8}{9} = \dfrac{32}{36}$ and $\dfrac{3}{4} = \dfrac{27}{36}$

Remember to include the whole numbers before adding. $2\dfrac{32}{36} + 3\dfrac{27}{36}$

The sum of the whole numbers is 5, and the sum of the fractions is $\dfrac{59}{36}$. So the sum of the mixed numbers is $5\dfrac{59}{36}$.

This simplifies to $6\dfrac{23}{36}$.

When we subtract mixed numbers, combine all terms and simplify the sum.

Study This Example

Subtract: $2\dfrac{8}{9}$ from $3\dfrac{3}{4}$. This is the same as $3\dfrac{3}{4} - 2\dfrac{8}{9}$.

First, the LCM of 9 and 4 is 36. So the common denominator is 36.

Determine equivalent fractions when the denominator is 36.

$\dfrac{3}{4} = \dfrac{27}{36}$ and $\dfrac{8}{9} = \dfrac{32}{36}$

Remember to include the whole numbers before subtracting $3\dfrac{27}{36} - 2\dfrac{32}{36}$

When we try to subtract our fractions, we can't take 32 from 27. So we have to regroup and rename $3\dfrac{27}{36}$ as $2\dfrac{63}{36}$.

Now we can subtract $2\dfrac{63}{36} - 2\dfrac{32}{36}$. The difference of these mixed numbers is $\dfrac{31}{36}$.

Try This

Add or subtract these mixed fractions, and simplify your answers.

1. $2\dfrac{5}{8} + 4\dfrac{3}{16}$

2. $6\dfrac{4}{5} - 3\dfrac{3}{7}$

3. $3\dfrac{8}{9} + 7\dfrac{1}{6}$

4. $4\dfrac{7}{12} - 2\dfrac{4}{15}$

5. $5\dfrac{7}{8} + 1\dfrac{3}{10}$

6. $4\dfrac{11}{16} - 3\dfrac{2}{3}$

The answers appear on page 210.

MULTIPLYING FRACTIONS

We can multiply fractions easily by multiplying the numerators and denominators separately and then simplifying the product.

Study This Example

$$\frac{4}{5} \times \frac{3}{8} = \frac{12}{40} = \frac{3}{10}$$

In multiplication, we can also simplify our fractions before multiplying.

$$\frac{{}^{1}\!\!\not{4}}{5} \times \frac{3}{\not{8}_{2}} = \frac{1}{5} \times \frac{3}{2} = \frac{3}{10}$$

Try This

Multiply these fractions, and simplify your answers.

1. $\dfrac{5}{8} \times \dfrac{3}{16} =$

2. $\dfrac{3}{7} \times \dfrac{4}{5} =$

3. $\dfrac{1}{6} \times \dfrac{8}{9} =$

4. $\dfrac{7}{12} \times \dfrac{4}{15} =$

5. $\dfrac{3}{10} \times \dfrac{7}{8} =$

6. $\dfrac{2}{3} \times \dfrac{11}{16} =$

The answers appear on page 211.

DIVIDING FRACTIONS

We can divide fractions by multiplying the first term by the **reciprocal** of the second. The reciprocal of a fraction is its multiplicative inverse, or the number that we multiply it by to get 1 as the product. For example, the reciprocal of $\dfrac{3}{4}$ is $\dfrac{4}{3}$. If we multiply these two fractions together, the product is 1. Notice that the numerators and denominators of reciprocals are inverted. In other words, what was the numerator is now below the fraction bar and what was the denominator is now on top in the reciprocal.

Study This Example

$$\frac{4}{5} \div \frac{3}{8}$$

We can rewrite this to multiply the $\frac{4}{5}$ by the reciprocal of $\frac{3}{8}$, which is $\frac{8}{3}$.

Our new problem is $\frac{4}{5} \times \frac{8}{3}$.

When we multiply the fractions, the product is $\frac{32}{15}$.

When we simplify, the quotient is $2\frac{2}{15}$.

In division, we can also simplify our fractions before multiplying them together. So $\frac{{}^1\cancel{4}}{5} \times \frac{3}{\cancel{8}_2} = \frac{1}{5} \times \frac{3}{2} = \frac{3}{10}$.

Try This

Divide these fractions, and simplify your answers.

1. $\frac{5}{8} \div \frac{3}{16} =$

2. $\frac{3}{7} \div \frac{4}{5} =$

3. $\frac{1}{6} \div \frac{8}{9} =$

4. $\frac{7}{12} \div \frac{4}{15} =$

5. $\frac{3}{10} \div \frac{7}{8} =$

6. $\frac{2}{3} \div \frac{11}{16} =$

The answers appear on page 211.

MULTIPLICATION AND DIVISION WITH MIXED FRACTIONS

As we know, mixed fractions are made up of both whole numbers and fractions. For example, $9\frac{1}{8}$ and $6\frac{2}{5}$ are mixed fractions that are sometimes called mixed numbers.

When we multiply and divide mixed numbers, we must change them to improper fractions before operating on them. Always remember to simplify the product or sum.

Study This Example

Multiply: $2\frac{8}{9} \times 3\frac{3}{4}$

First convert the mixed numbers to improper fractions:
$\frac{26}{9} \times \frac{15}{4}$

Next you can simplify the fractions: $\frac{13}{3} \times \frac{5}{2}$

Now you can multiply the numerators together and the denominators together to get the product $\frac{65}{6}$.

This simplifies to $10\frac{5}{6}$.

When we divide mixed numbers, we must change them to improper fractions before operating on them.

Study This Example

Divide: $2\frac{8}{9} \div 3\frac{3}{4}$

First convert the mixed numbers to improper fractions:
$\frac{26}{9} \div \frac{15}{4}$

Next rewrite the problem to multiply by the reciprocal: $\dfrac{26}{9} \times \dfrac{4}{15}$

Check to see if you can simplify the fractions: $\dfrac{26}{9} \times \dfrac{4}{15}$

They cannot be simplified. So multiply the numerators together and the denominators together.

The answer to the problem is $\dfrac{104}{135}$.

This fraction cannot be simplified. So $\dfrac{104}{135}$ is the final answer.

Try This

Multiply or divide these mixed fractions, and simplify your answers.

1. $2\dfrac{5}{8} \times 4\dfrac{3}{16}$

2. $6\dfrac{4}{5} \div 3\dfrac{3}{7}$

3. $3\dfrac{8}{9} \times 7\dfrac{1}{6}$

4. $4\dfrac{7}{12} \div 2\dfrac{4}{15}$

5. $5\dfrac{7}{8} \times 1\dfrac{3}{10}$

6. $4\dfrac{11}{16} \div 3\dfrac{2}{3}$

The answers appear on page 212.

LESSON 3—DECIMAL PLACE VALUE AND COMPUTATION

The set of decimals includes all numbers between zero and one. Decimals are named by the place value of their final digit. For example, we name the number zero and seven hundred twenty-four thousandths (0.724) that because the final digit, four (4), is in the thousandths place. Notice where each of the digits are located in the place value table.

PLACE VALUE TABLE FOR DECIMALS

Hundreds	Tens	Ones	Tenths	Hundredths	Thousandths	Ten Thousandths	Hundred Thousandths	Millionths
		0	.7	2	4			
5	6	4	.0	9	7	2	8	3

Think About This

Use the number 564.097283 in the place value table for decimals.

1. What is the value of the 3? _____

2. What is the value of the 5? _____

3. What is the value of the digit in the thousandths column? _____

4. What word separates the whole numbers from the decimals? _____

5. What symbol separates the whole numbers from the decimals? _____

The answers appear on page 212.

Try This

Multiple-Choice Questions

Use the decimal place value table for decimals if necessary.

1. In the number 148.96503, the value of the digit 5 is

   A. 5 hundred

   B. 5 thousand

   C. 5 hundredths

   D. 5 thousandths

2. In the number 3.4598, the 8 is in the

   A. tenths column

   B. hundredths column

   C. thousandths column

   D. ten-thousandths column

3. Which number is equivalent to 0.20050?

   A. two-hundred fifty ten-thousandths

   B. twenty thousand fifty

   C. two thousand five ten thousandths

   D. zero and two hundredths

4. What digit is in the tenths place in the number 1.0306?

   A. 0

   B. 1

   C. 3

   D. 6

Open-Ended Questions

1. Explain why the number 74 has no decimal point.

2. When would you use a decimal point for the number 74 and where would it be placed?

The answers appear on pages 212–213.

ADDING DECIMALS

Adding decimals is very similar to adding whole numbers. It is important to keep all of the digits in the correct columns according to place value. You must be especially careful to keep all decimal points aligned. You may see decimal addition problems written horizontally. It is helpful to rewrite them vertically before solving.

Study This Example

Add 39.876 + 263.28. Begin by rewriting the problem:

```
  39.876
+263.28
 303.156
```

Notice that the decimal point and all digits are in columns according to their place value.

Add the decimals as you would add whole numbers.

Try This

Add.

1. 756.298   2. 6,472.13   3.   1.92847
  + 34.12   +   47.87   +34.2534

4. 8,345.009 + 234.76

5. 0.0098 + 2.113

6. 10.000 + 0.34567

The answers appear on page 213.

SUBTRACTING DECIMALS

Subtracting decimals is very similar to subtracting whole numbers. It is important that you keep all of the digits in the correct columns according to place value. Be especially careful to keep all decimal points aligned. You may see decimal subtraction problems written horizontally. It is helpful to rewrite them vertically before solving.

Study This Example

Subtract 263.28 – 39.876. Begin by rewriting the problem:
$$
\begin{array}{r}
263.28 \\
-39.876 \\
\hline
223.404
\end{array}
$$

Notice that the decimal point and all digits are in columns according to their place value.

Include a zero as a place holder (example: 263.280) to make sure that you regroup correctly.

Try This

Subtract:

1. $\begin{array}{r} 756.298 \\ -34.12 \\ \hline \end{array}$ .2. $\begin{array}{r} 6,472.13 \\ -47.87 \\ \hline \end{array}$ 3. $\begin{array}{r} 211.92847 \\ -34.2534 \\ \hline \end{array}$

4. 8,345.009 – 234.76

5. 6.0098 – 2.113

6. 10.000 – 0.34567

The answers appear on page 213.

MULTIPLYING DECIMALS

Multiplying decimals is very similar to multiplying whole numbers. As with whole numbers, it is important that you align all of the digits in columns so you can operate accurately. The decimal points do not require their own column as in addition and subtraction. Instead, the decimal point separates the whole number and decimal value of each factor. You may see decimal multiplication problems written horizontally. It is helpful to rewrite them vertically before solving.

Study This Example

Multiply 23.2 × 3.98. Begin by rewriting the problem.

$$
\begin{array}{r}
23.2 \\
\times\ 3.98 \\
\hline
\end{array}
$$

Notice that the digits are in three columns but the decimal points are not aligned.

Now multiply as you would if they were whole numbers:

$$
\begin{array}{r}
23.2 \\
\times\ 3.98 \\
\hline
1856 \\
2088\ \ \\
696\ \ \ \\
\hline
\end{array}
$$

Include zero place holders if you need help to keep the digits aligned:

$$
\begin{array}{r}
1856 \\
20880 \\
69600 \\
\hline
92336
\end{array}
$$

The last step is to place the decimal point in the product. Since there was a total of three decimal places in the problem, there will be a total of three decimal places in the product. Decimal places come after the decimal point. So the final answer is 92.336.

Try This

Place the decimal points in the products.

1.	56.29	2.	72.1	3.	211.9
	×34.12		× 7.87		× 0.14
	19206148		567427		29666

Multiply.

4. 134.2	5.	5.76	6.	80.7
× 28		× 0.01		× 158

7. 8.345×23.76

8. 6.0098×2.1

9. 12×0.34567

The answers appear on page 214.

DIVIDING DECIMALS

The process of dividing decimals is very similar to dividing whole numbers. However, we cannot divide by a number that has a decimal point in it. So the first step in dividing by decimals is to transform the divisor into a whole number. We do this by multiplying by a multiple of 10 that will result in a whole number. For example, if the divisor is 12.3, we would multiply it by 10 to transform it to the whole number 123. The next step, which is very important, is to multiply the dividend by the same number. So if our dividend is 33.21, we would multiply it by 10 to transform it to 332.1. Then place the decimal point in the appropriate place in the quotient and divide as you would if you were working with whole numbers. You may see decimal division problems written horizontally. It is helpful to rewrite them vertically before solving.

Study This Example

Divide $23.01 \div 3.9$. Begin by rewriting the problem:
$$3.9\overline{)23.01}$$

Notice that the dividend has a decimal point. To transform this to a whole number, we multiply by 10. When we multiply the divisor by 10, we also multiply the dividend by 10.

We can rewrite the problem and place the decimal point

appropriately in the quotient: $39\overline{)230.1}$

Now divide as you would if they were whole numbers, using the steps

     Estimate and divide
     Multiply
     Subtract and compare
     Bring down and repeat:

$$
\begin{array}{r}
5.9 \\
39\overline{)230.1} \\
\underline{195} \\
35 \\
\underline{351} \\
0
\end{array}
$$

Try This

Place the decimal points in the quotients.

1. $5.44 \div 3.4 = 16$

2. $72.1 \div 1.4 = 515$

3. $211.8 \div 0.75 = 2824$

Divide.

4. $28\overline{)134.4}$

5. $0.01\overline{)5.76}$

6. $80.5\overline{)169.05}$

7. $834.5 \div 2.5$

8. $6.006 \div 0.21$

9. $12 \div 0.003$

The answers appear on page 214.

LESSON 4—PERCENTS

Another way of representing fractions and decimals is as percents. Percent means hundredth. Ten percent is written 10%. It has the value 0.10 or zero and ten hundredths.

We can also write 10% in the form of the fraction $\dfrac{10}{100}$, where 10 is the numerator and 100 is the denominator. Remember that the fraction $\dfrac{10}{100}$ can be simplified to $\dfrac{1}{10}$.

Let's look at another example. 5% is the same as five hundredths. We write that value as 0.05. When we write 0.05 as a fraction, the numerator is 5 and the denominator is 100, so the fraction is written as $\dfrac{5}{100}$. This simplifies to $\dfrac{1}{20}$.

Think About This

Complete this table.

Percent	Decimal	Fraction as Hundredths	Fraction in Simplest Form
		$\dfrac{50}{100}$	
15%			
	0.56		
		$\dfrac{28}{100}$	
	0.85		
			$\dfrac{3}{4}$
	0.25		
		$\dfrac{40}{100}$	
100%			1

The answers appear on page 215.

LESSON 5—INTEGERS

The set of integers includes all whole numbers
(0, 1, 2, 3 . . .) and their opposites, or negative values
(–3, –2, –1, 0 . . .). Let's look at the number line to
illustrate this point.

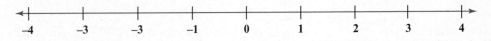

Zero is in the center, and the *positive* integers 1, 2, 3, and 4 *increase* (get larger) as they move further from zero to the right. The numbers to the left of zero are *negative* integers and their value *decreases* (gets smaller) as they move further from zero. Positive and negative values of the same number are called **opposites** or **additive inverses**. They are the same distance from zero, and each time we add or combine them the result is zero.

For Example

The numbers 5, 27, 999, and 1,234 are positive integers. They are greater than zero and are to the right of zero on the number line. Sometimes positive integers are written with a symbol to indicate that they are positive: +5, +27, +999, +1,234.

The numbers –9, –35, –671, and –2,487 are negative integers. They are less than zero and are to the left of zero on the number line. Negative integers are *always* written with a negative sign (–) to indicate that they are negative.

Zero is neither positive nor negative.

Try This

Multiple-Choice Questions

1. Which integer is the greatest?

   A. –2

   B. 0

   C. 36

   D. –121

2. Which integer is not positive or negative?

   A. –5

   B. 0

   C. 5

   D. –10

3. Which integer is the opposite of –25?

   A. $\dfrac{1}{25}$

   B. 0

   C. 0.25

   D. 25

4. Fill in the correct symbol –4 _____ –8

   A. <

   B. >

   C. =

   D. ≈

Extended Constructed-Response Question

Draw a number line and enter the numbers 0, 1, –1, 2, –2, 3, –3.

Short Constructed-Response Question

The temperature on a very cold day in January was 6°F below zero. How would you write this as an integer?

The answers appear on page 216.

ABSOLUTE VALUE

We have seen positive numbers that are greater than zero and negative numbers that are less than zero. Sometimes we evaluate both sets of numbers by their distance from zero on the number line. We call this absolute value. The absolute value of a number is its distance from zero on a number line. For example, +3 is 3 units from zero on the number line, so its absolute value is 3. Negative three (−3) is also 3 units from zero on the number line, so its absolute value is also 3. We use the symbol | | for absolute value. So |3| = 3 is read as "the absolute value of 3 is equal to 3." Additionally, |−3| = 3 is read as "the absolute value of negative 3 is equal to 3."

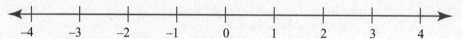

The absolute value of a number is never negative.

Try This

Find the absolute value of each number.

1. |15|

2. |−8|

3. |−27|

4. |13|

5. |−4|

6. |0|

7. Scientists recently discovered an alternative energy source 2,800 feet below sea level. Write this value as an integer.

8. What is the absolute value of the integer in question 7?

The answers appear on page 216.

LESSON 6—RATIONAL NUMBERS

We have worked with whole numbers, fractions, decimals and integers. All of these numbers belong to the set of rational numbers. The word "rational" stems from the word "ratio." Remember that a ratio is a relationship between two values.

Look at these examples: The whole number 5 is a rational number. It is the ratio of $\frac{5}{1}$ or 5:1. The fraction $\frac{7}{8}$ is a rational number. It is the ratio of 7 to 8 or 7:8. The decimal 0.38 is a rational number. It is the ratio of 38 to 100 or 38:100. The integer −99 is a rational number. It is the ratio of −99 to 1 or −99:1. The set of rational numbers also contains the opposites of fractions and decimals. All rational numbers have a location on the number line.

Try This

Study this number line.

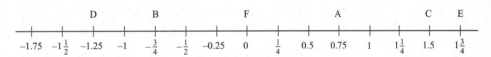

Identify which rational number corresponds with each point.

1. A

2. B

3. C

4. D

5. E

6. F

We can use the number line to compare rational numbers using <, >, or =.

Try This

Compare these rational numbers using <, >, or =.

1. –7 _____ 0

2. $\dfrac{5}{6}$ _____ –1

3. –2 _____ 0.75

4. $\dfrac{3}{8}$ _____ $-\dfrac{3}{8}$

5. 0 _____ –0.25

6. $-\dfrac{1}{2}$ _____ $-\dfrac{1}{4}$

The answers appear on page 217.

RATIONAL NUMBERS ON THE COORDINATE GRAPH

Now that we have worked with the set of rational numbers on a number line, the next step is to see them on a coordinate graph or, as it is sometimes called, on a coordinate grid. Think back to when you first plotted points on a coordinate graph. Remember that a coordinate graph has an *x*-axis (horizontal) and a *y*-axis (vertical). Also remember that each point on the graph is named by an ordered pair.

The ordered pair (0, 0) is the origin. To locate point *A* with the coordinates (3, 4), begin at the origin and move to the right along the *x*-axis 3 spaces and up the *y*-axis 4 spaces.

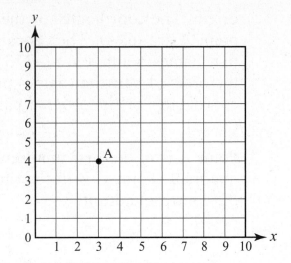

Try This

Plot the following points.

1. *B* (6, 3)

2. *C* (0, 4)

3. *D* (5, 1)

4. *E* (6, 3)

The answers appear on page 218.

Four-Quadrant Graph

Now let's look at a four-quadrant graph. It is four times as large as the graphs you have worked with before. It is expanded so we can plot positive and negative points. Notice the *x*-axis. It looks like the number line that we used with positive and negative integers. Notice that the *y*-axis is also extended to include negative numbers. When we plot the points named by our ordered pairs on the four-quadrant grid, positive *x*-coordinates are to the right of the origin and negative *x*-coordinates are to the left of the origin. Positive *y*-coordinates extend up from the origin, and negative *y*-coordinates extend below the origin. The coordinates or the ordered pair that names point *A* are (6, 4). The coordinates or the ordered pair that names point *B* are (–2, 3). The coordinates or the ordered pair that names point *C* is (–5, –4). The coordinates or the ordered pair that names point *D* is (7, –5).

Point *A* is in the first quadrant. Point *B* is in the second quadrant. Point *C* is in the third quadrant. Point *D* is in the fourth quadrant.

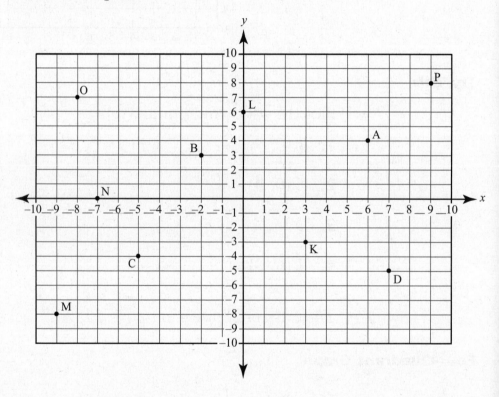

Try This

Plot these points on the four quadrant grid.

1. *E* (–10, 9)

2. *F* (0, –4)

3. *G* (9, –8)

4. *H* (–5, –5)

5. *I* (7, 0)

6. *J* (–6, 10)

Name the coordinates of each point.

7. *K*

8. *L*

9. *M*

10. *N*

11. *O*

12. *P*

The answers appear on pages 218–219.

GEOMETRY

LESSON 1—GEOMETRY REVIEW

Let's begin by reviewing how many of the geometric properties you remember.

1. Which of the following is the name for •A?

    A. line A

    B. point A

    C. ray A

    D. angle A

2. Which of the following names this figure?

    A. triangle

    B. pentagon

    C. hexagon

    D. square

3. Which of the following describes a line that goes up and down?

   A. horizontal

   B. straight

   C. parallel

   D. vertical

4. Which of the following is *not* a quadrilateral?

   A. octagon

   B. rhombus

   C. parallelogram

   D. trapezoid

5. Which of the following names the line in this circle?

   A. angle

   B. diameter

   C. circumference

   D. radius

6. Which of the following does *not* name a type of angle?

   A. acute

   B. right

   C. obtuse

   D. segment

7. Which of the following describes two geometric figures that are the same size and shape?

   A. congruent

   B. equilateral

   C. similar

   D. symmetry

8. Which of the following describes two geometric figures that have the same shape but are different sizes?

   A. congruent

   B. equilateral

   C. similar

   D. symmetry

9. Which of the following figures has six faces?

   A. cube

   B. cylinder

   C. circle

   D. cone

The answers appear on page 219.

LINES

Before we look at specific lines, let's review these symbols:

Point A: •A

Ray AB: $\overrightarrow{AB}$

Line segment AB: $\overline{AB}$

Line AB: $\overleftrightarrow{AB}$

Let's look at these examples of lines:

Parallel lines are lines that never cross or intersect:

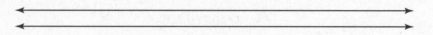

Perpendicular lines are lines that intersect to form 90 degree angles, or right angles:

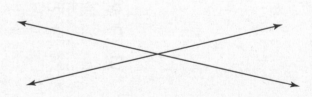

Intersecting lines are lines that cross at a given point:

Try This

Tell whether each pair of lines is parallel, perpendicular, or intersecting.

1.

2.

3.

4.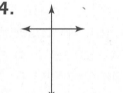

5.

6.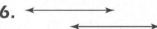

The answers appear on pages 219–220.

TRIANGLES

Triangles are formed when three line segments meet and form three angles.

Look at these examples:

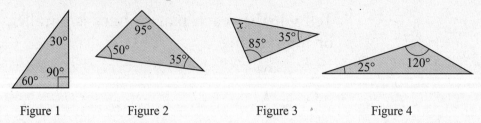

Figure 1   Figure 2   Figure 3   Figure 4

Each of these triangles has different angle measures but the sum of the angles of any triangle will always equal 180 degrees.

In the first example, the angles measure 30 degrees, 90 degrees, and 60 degrees. When you add them they total 180 degrees.

In the second example, the angles measure 95 degrees, 35 degrees, and 50 degrees. Their sum is also 180 degrees.

In the third example, one angle measures 35 degrees and another measures 85 degrees. To find the measure of the third angle, first add 35 and 85 together. This gives a total of 120 degrees. If we know that the total must equal 180 degrees, we can subtract to find the difference. 180 − 120 = 60 degrees. Check to make sure that 35 + 85 + 60 = 180.

In the fourth example, one angle measures 120 degrees and another measures 25 degrees. What is the measure of the third angle?

120 + 25 + _____ = 180. We know that 120 + 25 = 145. When we subtract 145 from 180, we know that the third angle is 35 degrees. Check to prove that 120 + 25 + 35 = 180.

Try This

Multiple-Choice Questions

1. Which of the following is true about triangles?

   A. they have three sides and three angles

   B. they are made up of three connecting rays

   C. they have two right angles

   D. none of the sides are the same

2. What does the sum of the angles in every triangle equal?

   A. 90 degrees

   B. it depends on the size of the triangle

   C. 180 degrees

   D. none of the above

3. One of the angles in a triangle measures 80 degrees and another measures 40 degrees. What is the measure of the third angle?

   A. 40 degrees

   B. 60 degrees

   C. 80 degrees

   D. 180 degrees

4. Two of the angles in a triangle each measure 45 degrees. What is the measure of the third angle?

   A. 45 degrees

   B. 90 degrees

   C. 135 degrees

   D. 180 degrees

Open-Ended Question

Kira says that the angles in her triangle measure 60, 70, and 80 degrees. Tommy says that's impossible. Who is right? Explain your answer.

The answers appear on page 220.

CLASSIFYING POLYGONS

Polygons are classified by their number of sides and angles. They are considered to be *regular* if all sides and angles are equal or congruent.

The sides of regular polygons are congruent. The angles of regular polygons are congruent.

Think About This

How many sides and angles does each polygon have?

1. triangle　　　＿＿＿＿＿

2. quadrilateral　＿＿＿＿＿

3. pentagon　　　＿＿＿＿＿

4. hexagon　　　＿＿＿＿＿

5. heptagon　　　＿＿＿＿＿

6. octagon　　　＿＿＿＿＿

7. nonagon　　　＿＿＿＿＿

8. decagon　　　＿＿＿＿＿

9. When is a rectangle a regular polygon?

The answers appear on page 220.

Two- and Three-Dimensional Shapes

Two-dimensional shapes are flat and are usually represented on a plane. Their two dimensions are length and width. All of the polygons that we just mentioned are two-dimensional shapes.

Think About This

1. What other two-dimensional shapes can you name?

Now look at these three-dimensional shapes. In addition to length and width, these figures have a third dimension—height. The surfaces of three-dimensional figures are called **faces**. Study these three-dimensional objects. Name them and tell how many faces each has.

2.

3.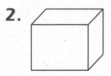

4.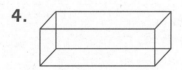

5. Can you name any other three-dimensional shapes?

The answers appear on page 221.

LESSON 2—UNITS OF MEASUREMENT

Let's begin by reviewing the systems of measurement and all of the units of measurement that we have worked with so far.

UNITS OF MEASUREMENT

	U.S. Customary	Metric
Length	Inch Fraction of an inch ($\frac{1}{8}, \frac{1}{4}, \frac{1}{2}$) Foot = 12 inches Yard = 3 feet Mile = 1,760 yards or 5,280 feet	Millimeter = 0.001 meter Centimeter = 0.01 meter Decimeter = 0.1 meter Meter Kilometer = 1,000 meters 1 Kilometer = 0.6 Mile
Weight	Ounce Pound = 16 ounces	Gram Milligram = 0.001 grams Kilogram = 1,000 grams
Capacity	Fluid Ounce Cup = 8 ounces Pint = 2 cups Quart = 2 pints or 4 cups Gallon = 4 quarts	Liter Milliliter = .001 liter
Time	Second Minute = 60 seconds Hour = 60 minutes Day = 24 hours Week = 7 days Month = 28, 30, or 31 days Year = 365 days (leap year = 366 days)	
Temperature	Degrees Fahrenheit—(water freezes @ 32 degrees, boils @ 212 degrees) Degrees Celsius—(water freezes @ 0 degrees, boils @ 100 degrees)	
Area	Square inch, square feet, square yards, square miles	Square millimeter, square centimeter, square decimeter, square meter, square kilometer
Volume	Cubic inch, cubic feet, cubic yards, cubic miles	Cubic millimeter, cubic centimeter, cubic decimeter, cubic meter, cubic kilometer

Try This

Multiple-Choice Questions

Refer to the table on page 64 to answer the following:

1. Rachel used three strips of tape to hang her poster. The strips measured 8 inches, 7 inches, and 6 inches. How much tape did Rachel use altogether?

   A. 20 inches

   B. 1 foot 9 inches

   C. 2 feet

   D. 1 yard

2. The distance from Julian's bedroom to the kitchen is 8 meters. He wants to extend his speakers to the kitchen. How much wire does Julian need?

   A. 800 millimeters

   B. 800 decimeters

   C. 800 centimeters

   D. 800 kilometers

3. Nicole is serving lemonade at her party. She expects that 12 people will each drink about 10 ounces. About how much lemonade should Nicole make?

   A. about 10 cups

   B. about 3 quarts

   C. about 6 pints

   D. about 1 gallon

4. Wilma was born on January 6 and Jason was born on May 25 of the same year. How much older than Jason is Wilma?

   A. 4 months, 2 weeks, and 5 days

   B. 7 months, 3 weeks, and 2 days

   C. 135 days

   D. 19 weeks

5. Calvin placed second in a 7-kilometer race. About how many miles did Calvin run?

   A. about 2 miles

   B. about 3 miles

   C. about 4 miles

   D. about 5 miles

6. The distance from Luis's house to the library is 8,500 meters. How many kilometers is that?

   A. 8.5 kilometers

   B. 850 kilometers

   C. 5,100 kilometers

   D. 8,500,000 kilometers

7. The temperature was 52 degrees Fahrenheit at 8:00 A.M. By noon, it was 72 degrees. What was the average temperature increase per hour?

   A. 5 degrees

   B. 10 degrees

   C. 15 degrees

   D. 20 degrees

8. Beth is bringing three pounds of fudge to the school picnic. She already made two and one-fourth pounds. How many more ounces of fudge does Beth need?

A. 10 ounces

B. 12 ounces

C. 16 ounces

D. 48 ounces

9. Hector is conducting a science experiment that requires one kilogram of topsoil. He has already weighed out 765 grams. How much more topsoil does Hector need?

A. 175 grams

B. 235 grams

C. 1,000 grams

D. .225 kilograms

10. Tai can run one lap in 90 seconds. How many laps can Tai run in 6 minutes?

A. 2

B. 4

C. 6

D. 8

Open-Ended Question

Max needs a box to store his trading cards. One box is a cube with faces that measure 6 inches. The other is a rectangular box that is 5 inches long, 4 inches wide, and 10 inches deep. Which box will hold more cards? Sketch and label each box and explain your answer.

The answers appear on page 221.

LESSON 3—MEASURING GEOMETRIC OBJECTS

PERIMETER

The distance around a polygon—a many-sided geometric figure—is called the **perimeter**. We solve for the perimeter by adding the lengths of the sides.

Look at these examples:

We add $14 + 6 + 14 + 6 = 40$ to solve for the perimeter of this rectangle. We label the perimeter in inches, the unit of measurement of each side.

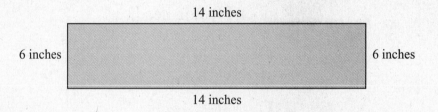

We add $14.6 + 9.2 + 7.5 = 31.3$ centimeters to solve for the perimeter of this triangle. We label the perimeter in centimeters, the unit of measure of each side.

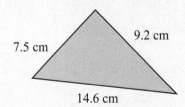

The perimeter of this trapezoid is 36 feet. What is the missing length?

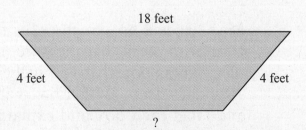

The missing side measures 10 feet. When we subtract the sum of the three sides (18 + 4 + 4 = 26) from 36, the perimeter, the result is 10.

The perimeter of all polygons is found by adding the lengths of all the sides.

CIRCUMFERENCE

The distance around a circle is called the **circumference**. We solve for the circumference by applying the formula $C = \pi d$. We read this as *circumference = pi times the diameter*. Pi (π) is a letter of the Greek alphabet. We use the symbol π to represent the value of the mixed decimal 3.14, or the improper fraction $\frac{22}{7}$. Most calculators have a π key that represents the value of π as 3.141592654. When we use a calculator, it is important to round our answer.

Remember that the diameter is twice the length of the radius, so to find the circumference, we can also apply the formula $C = 2\pi r$, (circumference = 2 × pi × radius), where we double the radius first and then multiply by 3.14, the value of π.

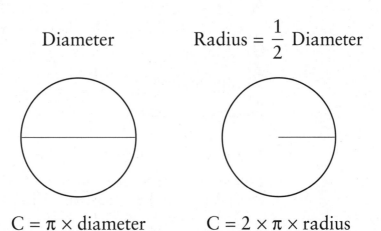

Diameter   Radius = $\frac{1}{2}$ Diameter

$C = \pi \times$ diameter $C = 2 \times \pi \times$ radius

Think About This

When would you use $\dfrac{22}{7}$ as π to solve for the circumference?

The answer appears on page 222.

Try This

Multiple-Choice Questions

1. What is the perimeter of a square with sides measuring 7.2 centimeters?

   A. 7.2 centimeters

   B. 14.4 centimeters

   C. 22.61 centimeters

   D. 28.8 centimeters

2. The length of a rectangle measures $6\dfrac{1}{2}$ inches and the width measures $5\dfrac{1}{4}$ inches. What is its perimeter?

   A. $11\dfrac{3}{4}$ inches

   B. 23.5 inches

   C. 26 inches

   D. 34.125 inches

3. The perimeter of a parallelogram measures 56 feet. Three of its sides measure 12 feet, 12 feet, and 16 feet. What is the measure of the fourth side?

   A. 12 feet

   B. 16 feet

   C. 28 feet

   D. 56 feet

4. The radius of a circular garden is 10.5 yards. What is its diameter?

   A. 5.25 yards

   B. 10.5 yards

   C. 21 yards

   D. 32.97 yards

5. What is the length of each side of an equilateral triangle that has a perimeter of 4 feet?

   A. 1 foot

   B. 12 inches

   C. 16 inches

   D. 48 inches

Extended Constructed-Response Question

The Wilsons installed a swimming pool that has a diameter of 16 feet in their square yard that measures 30 feet on each side. They want to install a fence around the swimming pool *and* around the entire yard. Draw a picture of the Wilson's yard and the swimming pool and tell how many total feet of fencing they will need. Show your work.

The answers appear on page 222.

AREA OF SQUARES AND RECTANGLES

Area is the number of square units needed to cover a figure, such as a polygon or circle. Look at this example:

This rectangle has a length of 5 units and a width of 7 units. Notice that when we fill the rectangle with square units, we are covering its area. There are 35 of them, or the area of the rectangle is 35 square units. We can also use the formula for finding the area of a rectangle that says *Area = length × width*. When we solve for the area of this rectangle by using the area formula, we substitute the values for the length and width, and multiply 5 × 7. The result is 35 *square* units.

The formula for finding the area of a square or rectangle is: Area = length × width

Substitute the values: Area = 5 × 7

Always remember to label your answer: Area = 35 square units

Try This

Sketch each rectangle, and find the area when:

1. Length = 52 cm   Width = 34 cm

2. Length = 26.5 ft   Width = 77.9 ft

3. Length = $7\frac{1}{2}$ mm   Width = $9\frac{1}{4}$ mm

4. Length = 125 mi   Width = 12.5 mi

5. Length = 0.23 cm   Width = 3.23 cm

6. Length = 96 mm   Width = $\frac{1}{8}$ mm

The answers appear on page 223.

Think About This

If multiplying length × width equals the area, you can solve for the length by dividing the area by the width. You can solve for the width by dividing the area by the length. So if l = 50 m and w = 12 m, A = 50 × 12 = 600 m^2. The length = 600 mm^2 ÷ 12 m = 50 m, and the width = 600 mm^2 ÷ 50 m = 12 m.

Try This

The area of a rectangle is 360 m^2. What is its length if:

1. w = 15 m

2. w = 1.2 m

3. w = $11\frac{1}{4}$ m

The area of a rectangle is 22.5 m^2. What is its width if:

4. l = 6 m

5. l = 0.225 m

6. l = 0.05 m

The answers appear on page 223.

AREA OF A PARALLELOGRAM

Finding the area of a parallelogram is similar to finding the area of squares and rectangles, although we use the formula *Area = base × height*.

For Example

Notice that this parallelogram does not have right angles like the rectangle, and that the sides are slanted. Rather than use the length of the sides to find its area, we use the lengths of the base and the height. Both are measured as the distance between the parallel sides. In this parallelogram, the base equals 8 feet and the height equals 3 feet.

$h = 3$ in.

$b = 8$ in.

We solve for the area by using the formula Area = base × height (A = bh).

Substitute the values: Area = 8 × 3

Remember to label with square feet: Area = 24 square feet

Try This

Sketch each parallelogram, and find the area when:

1. Base = 25 cm   Height = 43 cm

2. Base = 56.2 ft   Height = 79.7 ft

3. Base = $12\frac{1}{2}$ mm   Height = $8\frac{1}{4}$ mm

4. Base = 215 mi   Height = 21.5 mi

5. Base = 0.32 cm   Height = 2.23 cm

6. Base = 98 mm   Height = $\frac{1}{7}$ mm

The answers appear on page 223.

Think About This

If multiplying base × height equals the area of a parallelogram, you can solve for the base by dividing the area by the height. You can also solve for the height by dividing the area by the base. If b = 25 m and h = 6 m, A = 25 × 6 = 150 m². So b = 150 m² ÷ 6 m = 25 m, and h = 150 m² ÷ 25 m = 6 m.

Try This

The area of a parallelogram is 240 m². What is its base if:

1. h = 12 m

2. h = 1.6 m

3. h = $1\frac{1}{4}$ m

The area of a parallelogram is 12.5 m². What is its height if:

4. b = 5 m

5. b = 0.1 m

6. b = 0.05 m

The answers appear on page 224.

AREA OF A TRIANGLE

Once we are able to solve for the area of a parallelogram, it is easy to solve for the area of a triangle. Every parallelogram can be cut in half, or divided into two equal triangles. We can solve for the area of a triangle by multiplying its base times its height, and dividing the result by 2. We usually see that formula written as

$A = \frac{1}{2}$ bh, and we read it as, *area equals one-half base times height.*

Let's look at an example with a right triangle where the base equals 5 inches and the height equals 6 inches. Once we see that this triangle is one-half of a rectangle, we can solve for the area of the rectangle with a base or length of 5 inches and a height or width of 6 inches. Then we cut that value in half to determine the area of the triangle.

This is the same as using the formula area = $\frac{1}{2}$ base × height or $A = \frac{1}{2}$ bh.

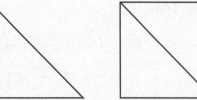

Area = (6 × 5) ÷ 2 = 15 in²
Area = 6 × 5 = 30 in²

Try This

Solve for the area of these right triangles.

1.

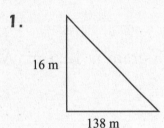

16 m

138 m

2.

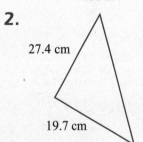

27.4 cm

19.7 cm

3.

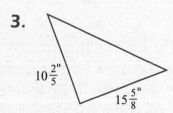

$10\frac{2}{5}"$

$15\frac{5}{8}"$

1. Base = _____   **2.** Base = _____   **3.** Base = _____

Height = _____   Height = _____   Height = _____

Area = _____   Area = _____   Area = _____

4. Base = 116 cm   **5.** Base = 23.4 cm   **6.** Base = $7\frac{1}{2}$ cm

Height = 84 cm   Height = 1.02 cm   Height = $2\frac{4}{9}$ cm

Area = _____   Area = _____   Area = _____

The answers appear on page 224.

Let's look at a way to solve for the area of a triangle that is *not* a right triangle. We will continue to use the formula area = $\frac{1}{2}$ base × height or A = $\frac{1}{2}$ bh. You can see from this example that the base of our triangle is 15 cm and one side measures 4 cm. It is important that you do not confuse this side length with the height. The height of the triangle is the distance from the base to the highest point of the figure. In this triangle, the vertical line drawn from the highest point to the base is its height. Here the height measures 3 cm. So to solve for the area of this triangle, we use the formula A = $\frac{1}{2}$ bh.

A = $\frac{1}{2}$ (15 × 3)

A = $\frac{1}{2}$ (45)

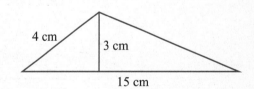

A = 22.5 cm^2

In another example, the base of the triangle is equal to 6.5 yards, and the height is equal to 9.2 yards.

Use the formula: $A = \dfrac{1}{2} bh$

$A = \dfrac{1}{2} (6.5 \times 9.2)$

$A = \dfrac{1}{2} (59.8)$

$A = 29.9 \text{ yd}^2$

The area is equal to 29.9 yd^2.

Try This

Solve for the area of these triangles.

1.

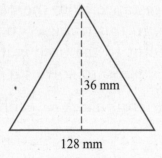

36 mm

128 mm

2.

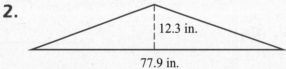

12.3 in.

77.9 in.

3.

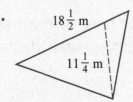

$18\frac{1}{2}$ m

$11\frac{1}{4}$ m

1. Base = _____   **2.** Base = _____   **3.** Base = _____

Height = _____   Height = _____   Height = _____

Area = _____   Area = _____   Area = _____

4. Base = 221 cm

Height = 175 cm

Area = _____

5. Base = 67.4 cm

Height = 3.05 cm

Area = _____

6. Base = $12\frac{1}{4}$ cm

Height = $3\frac{1}{7}$ cm

Area = _____

The answers appear on pages 224–225.

AREA OF SPECIAL QUADRILATERALS AND POLYGONS

Think About This

Which geometric figures make up this trapezoid?

What is the area of each geometric figure?

What is the area of the trapezoid?

You can solve for the area of almost any quadrilateral (four-sided figure) once you know how to apply the formulas for finding the area of a parallelogram and the area of a triangle. Let's look at this trapezoid again.

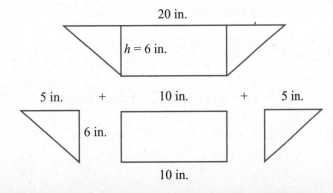

We can take our trapezoid and "decompose" it, or break it into shapes that we are more familiar with—a rectangle and 2 right triangles.

The base of the rectangle is 10 inches, and the base of each right triangle is 5 inches. The height of all figures that make up the trapezoid is 6 inches.

To solve for the area of the trapezoid, we can solve for the area of each figure separately and add them together.

The area of a rectangle is A = bh.

$$A = 10 \times 6$$

$$A = 60 \text{ in}^2$$

The area of the first triangle is A = $\frac{1}{2}$ bh.

$$A = \frac{1}{2}(5 \times 6)$$

$$A = 15 \text{ in}^2$$

The area of the second triangle is exactly the same as the first one because the bases and heights are the same.

$$A = \frac{1}{2}(5 \times 6)$$

$$A = 15 \text{ in}^2$$

Once we know the areas of all three parts of the trapezoid, we add them together to solve for the area of the entire figure.

$$60 + 15 + 15 = 90$$

The area of the trapezoid is 90 in^2.

Try This

Decompose these trapezoids, and find the area.

1.

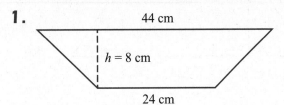

Area of rectangle _____

Area of first triangle _____

Area of second triangle _____

Area of trapezoid _____

2.

12 m   18 m

h = 9 m

Area of rectangle _____

Area of first triangle _____

Area of second triangle _____

Area of trapezoid _____

3.

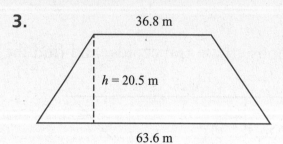

Area of rectangle   _____

Area of first triangle   _____

Area of second triangle   _____

Area of trapezoid   _____

The answers appear on page 225.

We can also solve for the area of a trapezoid by applying the formula Area = $\frac{1}{2}$ h(b_1 + b_2), where h is the height and where b_1 and b_2 are the lengths of the bases.

Try This

Use the formula A = $\frac{1}{2}$ h(b_1 + b_2) to solve for the area of these trapezoids.

1.   **2.**

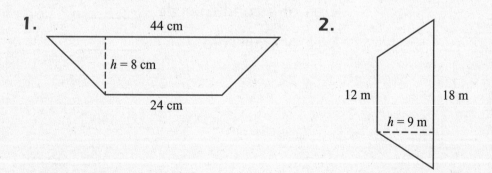

You should have the same results that you had when you decomposed each figure.

The answers appear on page 225.

Areas of Irregular Polygons

Find the area of each irregular polygon by decomposing them. ·

1.

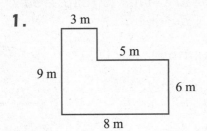

2.

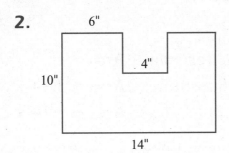

3.

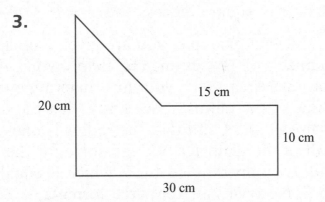

The answers appear on page 225.

AREA OF A CIRCLE

When we looked at the circumference of a circle, we reviewed the terms, pi(π), diameter, and radius. Remember that pi is equal to 3.14 or $\frac{22}{7}$, that the diameter is a chord that extends from one point on the circle and through the center to another point on its circumference, and that the radius is one-half the length of the diameter.

We solve for the area of a circle by squaring the radius, or multiplying it by itself, and then multiplying that answer by π. We write the formula for finding the area of a circle as Area = πr^2. This is read *area is equal to pi times the radius squared.*

Look at these examples:

This circle has a diameter of 12 inches and a radius of 6 inches.

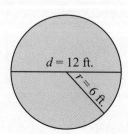

To find the area, we apply the formula: Area = πr^2

Substitute: Area = 3.14×6^2

Area = 3.14×36

Remember to label the answer in square inches: Area = 113.04 square inches

In the next example, you are given a circle with a diameter of 14 centimeters. Before you can apply the formula Area = πr^2, you must determine the value of the radius, r. We know that the diameter is 14 and that the radius is one-half of the diameter. We can solve for the radius by dividing 14 by 2, which is equal to 7. Now we can apply the formula

Area = πr^2. This time, use the value $\dfrac{22}{7}$ for pi because we are working with multiples of 7.

Start with the formula: Area = πr^2

Substitute: Area = $\dfrac{22}{7} \times 7 \times 7$

Area = $\dfrac{22}{7} \times 49$

Remember to label your answer square centimeters: Area = 154 square centimeters

Try This

Multiple-Choice Questions

1. A side of a square measures 22.8 centimeters. What is its area?

   A. 22.8 square centimeters

   B. 91.2 square centimeters

   C. 182.4 square centimeters

   D. 519.84 square centimeters

2. The area of a rectangle is 100 square inches. If the length is 25 inches, what is the width?

   A. 4 inches

   B. 400 square inches

   C. 225 square inches

   D. 2,500 inches

3. The base of a parallelogram is 16.5 centimeters and the height is 7.8 centimeters. Use the formula A = bh to solve for the area.

   A. 48.6 centimeters

   B. 66 square centimeters

   C. 128.7 square centimeters

   D. 404.12 square centimeters

4. A triangle is one-half of which figure?

   A. circle

   B. hexagon

   C. pentagon

   D. parallelogram

5. The base of a triangle is $2\dfrac{1}{4}$ inches and the height is $1\dfrac{1}{2}$ inches. Use the formula $A = \dfrac{1}{2}\,bh$ to solve for its area.

A. $\dfrac{15}{16}$ square inches

B. $1\dfrac{11}{16}$ square inches

C. $3\dfrac{3}{8}$ square inches

D. $7\dfrac{1}{2}$ square inches

6. Use the formula πr^2 to solve for the area of a circle with a radius of 10 centimeters.

   A. 3.14 square centimeters

   B. 31.4 square centimeters

   C. 314 square centimeters

   D. 3,140 square centimeters

7. Use the formula πr^2 to solve for the area of a circle with a diameter of 4 kilometers.

   A. 4 square kilometers

   B. 16 square kilometers

   C. 12.56 square kilometers

   D. 50.24 square kilometers

8. You want to cover a table that is 3 feet long and 4 feet wide. You should you use a tablecloth that measures

A. 7 square feet

B. 9 square feet

C. 11 square feet

D. 14 square feet

Open-Ended Question

Julius has a round picture frame with a radius of 4 inches. He wants to display a square photograph in it. Draw a picture of the frame and the photograph and give the length, width, and area of the photograph.

The answers appear on page 226.

SURFACE AREA

Think back to when you studied these three-dimensional figures—the cube, the cylinder, and the rectangular prism. Unlike polygons, which are flat, three-dimensional figures have many surfaces, or faces. The cube and the rectangular prism each have six faces, and the cylinder has three.

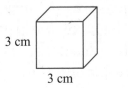

To determine the **surface area** of three-dimensional figures, we solve for the area of each face, and add the areas together.

Let's look at the cube.

Each face of a cube is a square. Suppose a side of the square measures 3 centimeters. We can use that information to solve for the area of one of the faces. We would multiply the length times the width, $3 \times 3 = 9$ square centimeters. Now that we know that the area of one face of the cube is 9 square centimeters and that the cube has six identical faces, we can solve for the total surface area by adding the areas of all six sides together, or multiplying the area, 9×6. The total surface area is equal to 54 square centimeters.

If we have a cube with a side that measures 7 yards, the area of one face is equal to 7×7, or 49 square yards, and the total surface area is 6 times 49, or 294 square yards.

We find the surface area of a rectangular prism by calculating the area of each of its faces and adding them together.

Notice that the top and bottom of a rectangular prism are identical.

The front is identical to the back. The right side is identical to the left side.

1. Label the rectangular prism above where $l = 4$ inches, $w = 3$ inches, and $h = 5$ inches.

Solve for the area of the top by multiplying 3 inches $\times$ 4 inches = 12 square inches. We know that if the top measures 12 square inches that the bottom does also. The area of the front measures 4 inches $\times$ 5 inches or 20 square inches. The back of the prism equals 20 square inches also. The left side measures 3 inches times 5 inches, or 15 square inches. This is true for the right side also.

When we add the measures of all six sides together, the total surface area is $12 + 12 + 20 + 20 + 15 + 15 = 94$ square inches.

Think About This

A rectangular prism's length = 10 feet, width = 9 feet, and height = 5 feet. If you need to, label this picture.

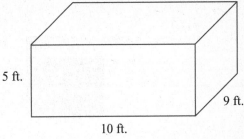

The length of the top measures 10 feet, and the width of the top measures 9 feet.

2. Give the area of the top.   _____

3. Give the area of the bottom.   _____

4. Give the area of the front.   _____

5. Give the area of the back.   _____

6. Give the area of the left side. _____

7. Give the area of the right side. _____

8. Add them up to determine the total surface area.   _____

The answers appear on pages 226–227.

USING NETS TO SOLVE FOR SURFACE AREA

Another way to solve for the surface area of three-dimensional figures is by using nets. A net is the flat pattern that you get if you open up a three-dimensional figure. For example, we know that a cube has six faces as seen in the first diagram. If we open and flatten the cube, it creates a net. The second diagram illustrates the net of a cube. We can fold the net back into its original shape and see it as a cube. If we unfold a rectangular prism, the result is a net that can be folded back into the original figure. The third diagram is a rectangular prism.

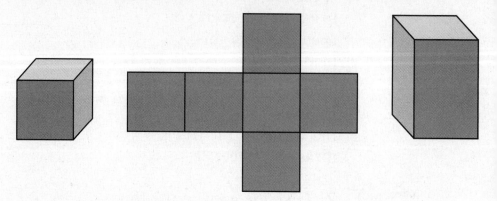

We can solve for the surface area of three-dimensional figures by representing them as nets, finding the area of each face, and adding the areas together.

Try This

1. Construct the net of a cube that has a side length of 1.5 inches. Solve for its surface area by finding the area of each of its faces and adding them together.

2. Construct a rectangular prism and its net where the length is 25 mm, the width is 30 mm, and the height is 35 mm. Solve for its surface area by finding the area of each face and adding the areas together.

The answers appear on page 227.

SURFACE AREA OF A CYLINDER

We follow the same process to find the surface area of a cylinder. A cylinder has three surfaces—two are circles and the third is a rectangle.

Look at this example:

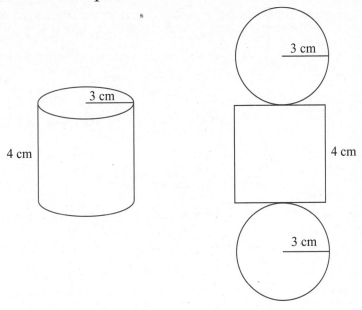

It is easy to recognize the top and bottom of this cylinder as two circles that are equal. To understand the third side, imagine slicing the cylinder open, and spreading it out flat.

The figure that appears is a rectangle. This is known as the **lateral surface** of the cylinder. When we solve for the area of each of the sides, we apply the formula for the area of a circle, $A = \pi r^2$, and the formula for the area of a rectangle, $A = l \times w$.

Let's try solving this example when the radius = 3 centimeters and the height = 4 centimeters.

Area of the top circle = πr^2

$$= 3.14 \times 3 \times 3$$

$$= 3.14 \times 9$$

$$= 28.26 \text{ square centimeters}$$

The area of the bottom circle is equal to the top circle or 28.26 square centimeters.

Now let's look at the rectangle. When we spread the cylinder out flat, the circumference of the circle becomes the length of the rectangle. The height of the cylinder becomes its width. But before we can apply the formula for finding the area of the rectangle, we have to solve for the circumference in order to determine its length. Remember the formula C = π × d, or 2 πr. If the radius is 3, we know that the diameter is 6. Now, substitute and solve C = 3.14 × 6 or 18.84 centimeters. Now that we know the length, we can apply the formula for the area of a rectangle and solve to find the lateral surface.

A = l × w

A = 18.84 × 4

A = 75.36 square centimeters

Let's put it all together now. The three sides of a cylinder that are added together to total the surface area are the two circles and the lateral surface, or the area of the rectangle. In this cylinder, the area of both of the circles is 28.26 square centimeters, and the area of our rectangle is 75.36 square centimeters. When we add them together, the surface area of the cylinder is 28.26 + 28.26 + 75.36 = 131.88 square centimeters. (Remember to always label your answer.)

Think About This

We want to solve for the surface area of a cylinder with a radius of 5 feet and a height of 8 feet. Use the formulas $A = \pi r^2$ and A = l × w.

1. How many surfaces does a cylinder have? _____

2. Name and number the surfaces. _____

_____

Substitute the correct values into the formula to solve for the area of the circles.

3. What is the area of the circles? ($A = \pi r^2$) _____

4. The length of the rectangle is equal to the
_____ of the circle.

5. The width of the rectangle = _____.

6. The circumference of the circle = _____ feet.

7. Therefore, the length of the rectangle = _____ feet.

8. The area of the rectangle or lateral surface =
_____ × _____ = _____.

9. The surface area of the cylinder =
_____ + _____ + _____ = _____.

10. What label will you give your answer?

_____

The answers appear on pages 227–228.

Try This

Multiple-Choice Questions

1. Which is *not* a dimension of a geometric solid, or prism?

    A. height

    B. length

    C. radius

    D. width

2. The area of which surface of a rectangular prism is equal to the area of the *top* of it?

    A. back

    B. bottom

    C. front

    D. left

3. What is the surface area of a rectangular prism where length = 9 meters, width = 7 meters, and height = 5 meters?

   A. 21 square meters

   B. 63.5 square meters

   C. 286 square meters

   D. 315 square meters

4. The base of a cylinder is a

   A. circle

   B. diameter

   C. radius

   D. rectangle

5. The lateral surface of a cylinder is a

   A. base

   B. circle

   C. circumference

   D. rectangle

6. The length of the lateral surface of a cylinder is equal to its

   A. area

   B. base

   C. circumference

   D. diameter

7. What is the surface area of a cylinder with a radius of 4 feet and a height of 3 feet?

A. 50.24 square feet

B. 75.36 square feet

C. 150.72 square feet

D. 175.84 square feet

Open-Ended Question

Erin is wrapping two gifts. One box is a cube where each side measures 8 inches. The other box has a length of 6 inches, a width of 8 inches, and a height of 10 inches. Which box requires more wrapping paper? Explain your answer.

The answers appear on page 228.

VOLUME

The term **volume** refers to the amount that can fit inside a three-dimensional figure, or how much a container can hold. To determine the volume, multiply the measures of each of the three edges. These lengths are also called the dimensions.

Look at this example:

The dimensions, or the length, width, and height of this rectangular prism, are 12 centimeters, 3 centimeters, and 2 centimeters.

3 cm

2 cm

12 cm

To solve for the volume,

Apply the formula: $V = l \times w \times h$

Substitute: $V = 12 \times 3 \times 2$

Solve: $V = 72$ cubic centimeters

Another way to look at solving for the volume of a prism is to solve for the area of the base, and then multiply that answer by the height. In this prism, the dimensions of the base are 12 centimeters and 3 centimeters. Area = l × w, or 12 × 3, or 36. When we multiply that result by the height, 2, we get the same volume, 72 cubic centimeters.

Try This

Solve for the volume of this prism by multiplying the measure of its edges or by using the formula $V = l \times w \times h$, when l = 7 inches, w = 6 inches, and h = 8 inches.

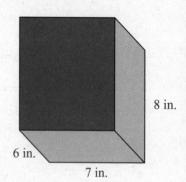

8 in.

6 in.

7 in.

1. Substitute: _____ × _____ × _____ = _____

2. How will you label your answer? _____

The answers appear on page 229.

Sometimes, the volume formula is written in the form $V = Bh$. Notice that the uppercase B is different from the lowercase b that we have been using as the variable for the base. The formula $V = Bh$ is read as "volume equals the area of the base times the height." To solve this, we solve for the area of the base of the rectangular prism first and then multiply that product by the height. When we apply this formula to the figure above, we see that B, or the area of the base, is equal to 42 square inches. When we multiply that value by the height of 8 inches, we see the volume is equal to 336 cubic inches.

Try This

Label each dimension, and solve for the area of the base and the volume of the following rectangular prisms. Use the formula V = l × w × h or V = Bh. You should get the same result. Remember to label your answers with the correct units.

1.

l = 28 mm

w = 24 mm

h = 136 mm

B = _____

V = _____

2.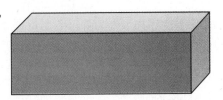

l = 55.5 in

w = 10.2 in

h = 8.8 in

B = _____

V = _____

3.

$l = 4\frac{5}{8}$ cm

$w = 3\frac{1}{2}$ cm

h = 10 cm

B = _____

V = _____

The answers appear on page 229.

VOLUME OF A CYLINDER

To find the volume of a cylinder, we first solve for the area of the base and then multiply it by the height.

Remember when we worked with surface area, we determined that the base of a cylinder is a circle and that the formula for finding the area of a circle is $A = \pi r^2$.

For Example

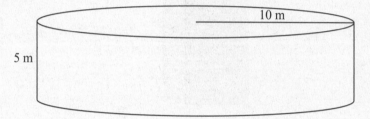

Solve for the volume of this cylinder, which has a radius of 10 meters and a height of 5 meters.

First, find the area of the base, which is a circle with a radius of 10 meters.

Use the formula: $A = \pi r^2$

Substitute: $A = 3.14 \times 10 \times 10$

Solve: $A = 314$ square meters

Now take the result of the area of the base and multiply it by the height of the cylinder to determine the volume.

We write this formula as $V = Bh$ and read it as *volume equals area of the base times height*. Notice that we use an uppercase "B" for the area of the base, unlike the lowercase "b" that we use for the base.

Apply the formula: $V = Bh$

Substitute the area of the base: $V = 314 \times 5$

Multiply to solve: $V = 1,570$ cubic meters

The volume of the cylinder is 1,570 cubic meters. (Remember to use the correct label.)

Think About This

1. The base of a cylinder is a _____.

2. The formula for finding the area of a circle is

   _____.

3. The formula $V = Bh$ is read _____

   _____.

4. The area of the base of a cylinder that has a radius of 7 inches and a height of 3 inches is _____.

5. The volume of that cylinder is _____.

6. We label volume in _____.

The answers appear on page 229.

Try This

Multiple-Choice Questions

1. What is the volume of a rectangular prism when the length = 3.5 feet, the width = 2.5 feet, and the height = 4 feet?

   A. 10 feet

   B. 35 cubic feet

   C. 48 cubic feet

   D. 56 cubic feet

2. The volume of a rectangular prism is 240 square meters. Its length is 10 meters and its width is 3 meters. What is its height?

   A. 8 meters

   B. 13 meters

   C. 24 meters

   D. 30 meters

3. What is the area of the base of a cylinder with a radius of 12 centimeters?

   A. 24 centimeters

   B. 75.36 square centimeters

   C. 144 square centimeters

   D. 452.16 square centimeters

4. The length of a side of a cube measures 1 foot. What is its volume?

   A. 1 foot

   B. 1 square foot

   C. 1 cubic foot

   D. 1 yard

5. What is the volume of a cylinder that has a radius of 1 meter and a height of 3 meters?

   A. 3.14 cubic meters

   B. 9.42 cubic meters

   C. 18.84 cubic meters

   D. 28.26 cubic meters

Open-Ended Question

Matt wants to save his sports trophies in the largest possible storage container. Should he buy a rectangular container that measures 4 feet × 5 feet × 6 feet, or a cube that measures 5 feet on each side? Which container should Matt buy? Explain your answer.

The answers appear on pages 229–230.

POLYGONS ON THE COORDINATE GRAPH

Sometimes we study two-dimensional figures or shapes on a coordinate grid or a graph. Remember that a coordinate graph has an *x*-axis (horizontal), and a *y*-axis (vertical), and that each point on the graph is named by an ordered pair. The ordered pair (0, 0) is the origin. To locate point A with the coordinates (3, 4), begin at the origin and move across the *x*-axis 3 spaces right and up 4. See the graph below for the other points that are plotted.

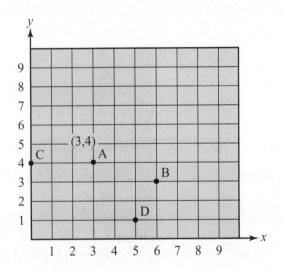

Think About This

1. What are they?

   B. (,)

   C. (,)

   D. (,)

2. Now connect point A to B, B to D, D to C, and C to A.

3. What figure did you make?

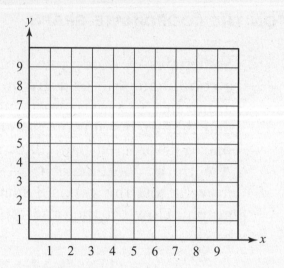

4. Plot four points on the coordinate grid that make a square when they are connected. List the ordered pairs of the square.

5. Begin at the coordinates (3, 2). Label two more points that can be connected to form rays that form a right angle.

6. Connect the points (1, 5) and (8, 5). Draw a line segment that is parallel to it and label the endpoints with ordered pairs.

The answers appear on pages 230–231.

EXPRESSIONS AND EQUATIONS

LESSON 1—PATTERNS, MODELS, AND GRAPHS

There are many different ways to work with expressions and equations. Recognizing, describing, and extending number patterns is a good way to begin.

Look at this example: 2, 4, 6, 8, . . . You should recognize and describe this pattern as counting by 2, or adding 2. We can extend it by including the numbers 10, 12, 14, and so on.

In this example: 100, 90, 80, 70, . . . the pattern is formed by subtracting 10 and we can extend it to include 60, 50, 40, 30 . . .

Numbers patterns are not limited to whole numbers. Think back to chapter 1, where we worked with fractions, decimals, and integers. Number patterns often include these sets of numbers also.

Look at these examples:

■ $\frac{1}{10}, \frac{2}{10}, \frac{3}{10}, \frac{4}{10}$ . . . Our pattern is to increase by $\frac{1}{10}$ or to add $\frac{1}{10}$. When we extend the pattern, the next fractions are $\frac{5}{10}, \frac{6}{10}, \frac{7}{10}$ . . .

■ 3.25, 3.50, 3.75, 4.0 . . . Our pattern is to increase by .25 or to add 25 hundredths. When we extend the pattern, the next decimals are 4.25, 4.50, 4.75 . . .

■ 1, 0, –1, –2 . . . Our pattern is to decrease by 1 or to subtract 1. When we extend the pattern, the next integers are –3, –4, –5 . . .

Try This

Multiple-Choice Questions

1. Name the next number in this pattern: 3, 8, 13, 18 . . .

   A. 4
   B. 16
   C. 19
   D. 23

2. Name the next number in this pattern: 5, $4\frac{1}{2}$, 4, $3\frac{1}{2}$ . . .

   A. 2
   B. 3
   C. $4\frac{1}{4}$
   D. $5\frac{1}{2}$

3. Name the next number in this pattern: 5.25, 5.35, 5.45, 5.55 . . .

   A. 5.56
   B. 5.60
   C. 5.65
   D. 6.00

4. Name the next number in this pattern: 0, –1, –2, –3 . . .

    A. –4

    B. –3

    C. –2

    D. –1

The answers appear on page 231.

Patterns can also be represented in tables. We usually describe them by using expressions or simple equations.

Look at this example. Notice that we begin by listing a series of numbers in one column that is labeled with the variable n. The second column describes the operation of adding 5 to each of the values of n, resulting in a pattern. We can extend this table by including additional values for n, and extending the pattern in the column for $n + 5$.

n	$n + 5$
1	6
3	8
5	10
7	12
9	14

A verbal rule that describes this pattern is add 5 to n.

A simple equation that describes this pattern is $n + 5$.

Each time we enter a different value for n, the result for the value of $n + 5$ changes also.

Think About This

1. What is the value of $n + 5$ when n is equal to 4?

n	2.0	3.0	4.0	5.0	6.0
$n - 0.8$	1.2	2.2	3.2	4.2	5.2

Sometimes tables are represented horizontally as in the example above.

A verbal rule that describes this pattern is subtract 0.8 from n.

A simple equation that describes this pattern is $n - 0.8$.

2. What is the value of $n - 0.8$ when n is equal to 6.6?

The answers appear on page 232.

Try This

Complete these tables:

1.

n	$n \times 5$
0	
3	
	35
	50
14	

2.

n	$1\frac{1}{8}$	$3\frac{3}{8}$			9
$n - \frac{1}{8}$			5	$6\frac{7}{8}$	

3. State a verbal rule that describes these patterns.

4. Give a simple equation that describes these patterns.

These answers appear on page 232.

TABLES

Patterns that are represented in tables and described by using simple equations are often referred to as **functions**. When we look at the examples that we've seen as functions, the values that we give to the variable n are called **input**, and the results, after operating on them are called **output**. The simple equation that describes the pattern is called the **rule**.

Functions can be represented as **verbal models, simple equations, tables,** or **graphs.**

Think about this example of a function:

$3n + 2$ is a rule in the form of a simple equation that represents the verbal models of the function: three times a number n plus two, or two more than three times a number n. We can set up an input/output table to show the input values of n and the output values of the function.

n	$3n + 2$
0	2
1	5
4	14
7	23

You should be able to solve for the output of a function when you know the rule and you should be able to state the rule when you know the input and output values of the function.

Study these function tables and state the rules.

1.

n	?
0	7
3	10
6	13
9	16

_____

2.

n	?
10	8
8	6
5	3
2	0

_____

3.

n	?
0	0
3	−3
6	−6
9	−9

_____

4.

n	?
0	1
3	7
6	13
9	19

_____

The answers appear on page 232.

MODELS

We used the variable n to represent the input in our function tables. Sometimes the rule and output are given and we must solve for n.

Look at this example.

You are given the equation $5 + n = 13$ where $5 + n$ is the rule and 13 is the output. You are asked to solve for n.

Thinking about this as a verbal model makes it easier to solve. Begin by asking yourself, "What number plus five is equal to 13?" You should recall that $5 + 8 = 13$, and therefore n is equal to 8.

You should be able to use variables to represent unknown quantities in both verbal models and equations. For example, two times the sum of 3 plus n is equal to 8 can also be written as $2 \times (3 + n) = 8$.

Five less than the product of ten times a number is equal to thirty-five can be written as $(10 \times n) - 5 = 35$.

Try This

Multiple-Choice Questions

Solve for n. Use a verbal model if necessary.

1. $3 \times n = 36$

    A. 3

    B. 9

    C. 12

    D. 36

2. Which equation represents the verbal model 2 more than a number is equal to 9?

   A. $2 + n = 9$

   B. $9 + n = 2$

   C. $2 + 9 = n$

   D. $2 \times n = 9$

3. What verbal model represents the equation $4 \times (3 + n) = 16$?

   A. Four times three times a number is equal to 16

   B. Four times three plus n is equal to sixteen

   C. Four times the sum of three plus n is equal to 16

   D. Four times three is equal to 16 plus three

4. What is the value of n in the equation $3 \times (n - 4) = 6$?

   A. 2

   B. 4

   C. 6

   D. 8

5. What is the value of n in the equation $(2 \times 4) - n = 4$?

   A. 2

   B. 4

   C. 6

   D. 8

The answers appear on page 233.

GRAPHS

Sometimes we use graphs to represent data. Let's think about this situation. A car travels for five hours at the rate of fifty miles per hour. We can enter this data in a function table where n is the number of hours and $n \times 50$ is our rule, which represents the total number of miles traveled.

n (hours)	$n \times 50$ (miles traveled)
1	50
2	100
3	150
4	200
5	250

Next, we can take this information and represent it on a graph. Think back to chapter 2 when we plotted points on a coordinate graph. Each of the values of n in the first column represent the x-coordinate, and each of the values of $n \times 50$ in the second column represent the y-coordinate.

The ordered pairs from this function table are (1, 50), (2, 100), (3, 150), (4, 200), and (5, 250).

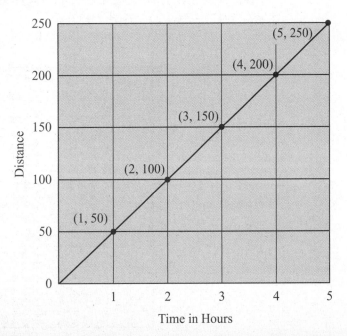

You've seen that when we work with patterns, functions, and relationships, we can model them in tables, as verbal models, in numerical expressions, equations, and on graphs. In most cases, the function is graphed as a line, and is therefore referred to as a **linear equation**. We solve linear equations by following the same procedure that we used when we worked with function tables.

Let's look at this example again.

n	$n + 5$
x	y
1	6
3	8
5	10
7	12
9	14

In our original example, we used the variable n to represent our input value. In a linear equation, we refer to this as our x-value. When we apply this to our function rule, we refer to the solution as our y-value.

The linear equation that represents this function is $x + 5 = y$, or $y = x + 5$. The solution for our linear equation is found in the second column of our table, or on the graph of a line containing the ordered pairs (1, 6), (3, 8), (5, 10), (7, 12), and (9, 14).

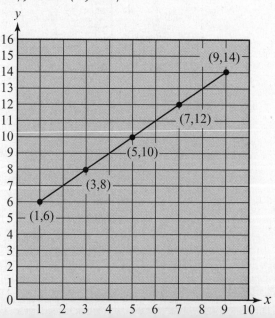

1. Solve the linear equation $y = 3x + 1$ when $x = 0, 1, 2,$ and 3. Show your solutions in a function table and on a graph.

x	$3x + 1$
0	
1	
2	
3	

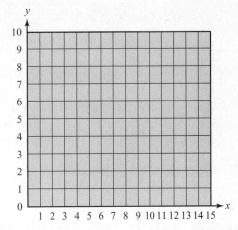

The answers appear on page 233.

LESSON 2—THE DISTRIBUTIVE PROPERTY

Solve $(4 \times 3) + (4 \times 8) =$ _____

Solve $4 \times (3 + 8) =$ _____

Did you get the answer 44 for both equations?

This is an example of the distributive property. The **distributive property** allows you to add (or subtract) the numbers in expressions first when you are multiplying them by a common factor.

Let's look at a few more examples:

$(5 \times 6) + (5 \times 4) =$
$\qquad 30 + 20 = 50$ and $5 \times (6 + 4) =$
$\qquad\qquad 5 \times 10 = 50$

$(6 \times 8) - (6 \times 3) =$ and $6 \times (8 - 3) =$
$\qquad 48 - 18 = 30$ $\qquad\qquad 6 \times 5 = 30$

$9 \times (7 - 4) =$ and $(9 \times 7) - (9 \times 4) =$
$\qquad 9 \times 3 = 27$ $\qquad\qquad 63 - 36 = 27$

Fill in the blanks:

$$(7 \times 6) + (7 \times 3) =$$

$$7 \times (\underline{\quad} + \underline{\quad}) =$$

$$7 \times \underline{\quad} = 63$$

$$(5 \times 7) - (5 \times 2) =$$

$$5 \times (\underline{\quad} - \underline{\quad}) =$$

$$5 \times \underline{\quad} = \underline{\quad}$$

The answers appear on page 233.

MORE ABOUT THE DISTRIBUTIVE PROPERTY

We can apply the distributive property when we study equivalent expressions. Look at the example $15 + 25 = 5(3 + 5)$. Think back to when we studied about common factors. Notice in this equation that we made an equivalent expression by distributing the greatest common factor, 5, over each of the terms. The solution remains the same. We can check this by solving each side of the equation. $15 + 25 = 40$ and $5 \times (3 + 5) = 5 \times 8 = 40$.

Here's another example:

$27 + 33 = 3(9 + 11)$ Distribute the GCF, 3.
$60 = 3(20)$ Solve the left side of the equation.
$60 = 60$ Solve the right side of the equation.

Notice that the answer is the same.

VARIABLES IN EXPRESSIONS

Sometimes a variable is used as an expression in an equation. Remember that a variable is a letter that takes the place of a number. When we see variables in expressions, we can operate on them like we do with numbers. Look at $n + n = 2n$. The variable, n, can stand for any number. We know from working with whole numbers that adding a number to itself is the same as multiplying it by 2. The same is true for variables. So when we add 2 n together, the result is $2n$. The same is true when we add 3 n. The result is $n + n + n = 3n$. Adding 4 n is $n + n + n + n = 4n$.

Try This

Multiple-Choice Questions

Use the distributive property to solve the following:

1. $6 \times (4 + 2) =$

    A. $(6 + 4) \times (6 + 2)$

    B. 6×6

    C. $(6 \times 4) \times (6 \times 2)$

    D. 6×8

2. $(8 \times 5) - (8 \times 3) =$

    A. 8×2

    B. $8 - 8$

    C. $8 - 2$

    D. 8×8

3. $(9 \times 3) + (9 \times 6) =$

    A. 27

    B. 54

    C. 72

    D. 81

4. $7 \times (8 - 2) =$

    A. 14

    B. 42

    C. 56

    D. 70

Short Constructed-Response Questions

5. $21 + 35 =$ _____ $(3 + 5)$

    _____ $=$ _____

6. $x + x + x + x + x + x =$ _____

The answers appear on page 234.

LESSON 3—EVALUATING EXPRESSIONS AND INEQUALITIES

EVALUATING EXPRESSONS

Remember that variables are letters or symbols that take the place of numbers in algebraic expressions or equations. We are often asked to evaluate expressions by giving values to the variables in them.

You may be asked to evaluate the expression $x + 5$ when x is equal to 6. The process we use to solve is to substitute and simplify. This means that we substitute the value 6 for x, and simplify, or solve for the answer. The solution is 11, because $6 + 5 = 11$.

Let's look at another example.

Evaluate the expression $10 - n$ when $n = 4$.

When we substitute 4 for n, we can solve the expression $10 - 4$ and the solution is 6.

Try This

Substitute and simplify to evaluate the following expressions:

1. $12 + n$ when $n = 8$

   A. 3

   B. 4

   C. 20

   D. 96

2. $15 \times n$ when $n = 3$

   A. 5

   B. 12

   C. 18

   D. 45

3. $32 \div n$ when $n = 8$

   A. 4

   B. 24

   C. 40

   D. 256

4. $125 - n$ when $n = 25$

   A. 5

   B. 100

   C. 150

   D. 225

5. $8 \times (n + 3)$ when $n = 2$

    A. 8

    B. 16

    C. 24

    D. 40

Short Constructed-Response Questions

6. Evaluate the expression n^3 when n is equal to 7.

7. Evlauate the equations $2n^2 - 5$ when n is equal to 3.

The answers appear on page 234.

INEQUALITIES

So far, the only expressions that we have worked with include values that are equal and are connected by an equal sign (=). Sometimes we work with expressions that are not equal and therefore use the symbol ≠, read as **not equal to.**

For example, $3 \neq 4$ is read three is not equal to 4, or $7 + 15 \neq 23$ is read seven plus fifteen is not equal to twenty-three.

Other inequality symbols that we use are ≤, **less than or equal to,** and ≥, **greater than or equal to.**

The expression $7 \leq 8$ is read seven is less than or equal to 8.

$5 + 4 \geq 9$ is read five plus four is greater than or equal to nine.

When we solve expressions having inequality symbols, there are many solutions. For example, in the inequality $n + 5 \neq 10$, the value of n is every number except 5, because 5 is the only number that when added to 5 will equal 10.

In the expression $n - 3 \geq 6$, the solution is 9 and all numbers greater than 9.

We can represent the solution on a number line. Notice that the number 9 is marked with a filled in circle and all points to the right of 9 are shaded as well.

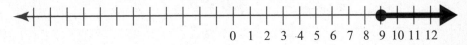

0 1 2 3 4 5 6 7 8 9 10 11 12

Try This

Multiple-Choice Questions

1. Which of these expressions is true?

    A. $6 + 3 = 10$

    B. $6 + 3 \leq 10$

    C. $6 + 3 \geq 10$

    D. $6 + 3 > 10$

2. Which of these is a value for n in the expression $n \leq 4$?

    A. 4

    B. 5

    C. 6

    D. 7

3. Which is *not* a solution for $n \neq 6$?

    A. $2 + 3 = n$

    B. $2 \times n = 12$

    C. $30 \div 3 = n$

    D. $3 + n = 8$

4. Which expression has the solution of all numbers ≥ 8?

   A. $1 + n \geq 10$

   B. $10 - 2 = n$

   C. $4 \times n \geq 32$

   D. $3 \times n \geq 27$

Short Constructed-Response Questions

5. Represent the solution for the expression $n < 0$ on a number line.

6. Represent the solution for the expression $n + 3 \geq 5$ on a number line.

Extended Constructed-Response Question

The Johnson family spent the day at the Slip-and-Slide Water Park. The admission charges are as follows:

Children ≤ 3 years old—Free

Students ≤ 12 years old—$12.00

Teens ≥ 13 years old—$18.00

Adults ≥ 18 years old—$24.00

What was the total admission fee for Mr. and Mrs. Johnson and their children:

Jerome—age 16

Luisa—age 13

William—age 12

Tina—age 3

Sammy—age 1

Show your work and explain your answer.

The answers appear on page 235.

STATISTICS AND PROBABILITY

LESSON 1—UNDERSTANDING DATA

Data is another word for information. Some of the places that we gather data is from research reports, newspapers, or by collecting it ourselves. One way to collect data is by conducting a survey or asking different people to respond to the same question. Once we have collected their responses, we can organize and display the data in different ways.

Let's look at this example:

The school's food service is changing their menu and wants to know what the students' favorite lunches are. They surveyed 100 sixth graders and asked them which they preferred—pizza, chicken nuggets, garden salad, tacos, hot dogs, or veggie burgers. When all of the data (information) was collected, they determined that 20 students chose pizza, 35 chose chicken nuggets, 10 chose garden salad, 12 chose tacos, 18 chose hot dogs, and 5 chose veggie burgers. Now that we have generated and collected the data, we can organize and display it in the following ways:

FREQUENCY TABLE

Food	# of Students
Pizza	20
Chicken Nuggets	35
Garden Salad	10
Tacos	12
Hot Dogs	18
Veggie Burgers	5

BAR GRAPH

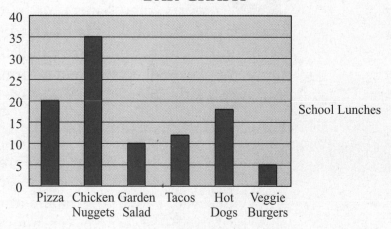

School Lunches

CIRCLE GRAPH

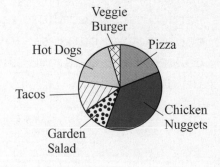

BAR GRAPHS

Notice that the data generated from our survey is displayed on different types of graphs. The bar graph represents each type of food on a vertical bar that extends to a height that is indicated on the left margin of the graph. By studying the bar graph, we can tell that each horizontal line represents five students. The height of the bar shows how many students selected each lunch choice. Notice that the order of each entry on the bar graph matches the order that it is presented in the table.

Try This .

Multiple-Choice Questions

This bar graph represents the number of containers of each type of juice that the juice company sent to the school. Use the data in this bar graph to answer the following questions:

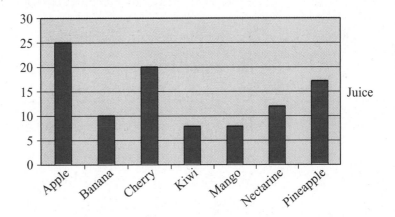

1. What is the total number of juices that were sent to the school?

   A. 70

   B. 80

   C. 90

   D. 100

2. How many more apple juices than cherry juices were sent?

   A. 5

   B. 13

   C. 25

   D. 45

3. What is the total number of apple, banana, and cherry juices?

   A. 10

   B. 20

   C. 25

   D. 55

4. Which three types of juice can be packed together in a carton that holds exactly 50 containers?

   A. apple, kiwi, and nectarine

   B. apple, cherry, and mango

   C. apple, mango, and pineapple

   D. apple, banana, and nectarine

5. Which two juices combined equals the number of containers of cherry juice?

   A. banana and kiwi

   B. kiwi and pineapple

   C. mango and nectarine

   D. mango and pineapple

Open-Ended Question

Construct a bar graph to represent five different amounts of different types of candy bars. Label your bar graph.

The answers appear on page 236.

CIRCLE GRAPHS

A **circle graph** resembles a pie and is sometimes referred to as a pie chart or pie graph. Each piece of data is represented as a slice of the pie. The greater the amount, the bigger the slice.

Think back to chapter 1 when we worked with percents and we said that 100 percent represents an entire amount. When we relate that concept to a circle graph, we see that each slice represents a percentage of the data collected. Remember the lunch survey that we worked with earlier.

Food	# of Students
Pizza	20
Chicken Nuggets	35
Garden Salad	10
Tacos	12
Hot Dogs	18
Veggie Burgers	5

Of the 100 students surveyed, 20 out of 100, or 20 percent, preferred pizza.

35 out of 100, or 35 percent, chose chicken nuggets.

Think about the percent of students who chose garden salad, tacos, hot dogs, and veggie burgers.

Once we determine the percent for each of the categories, we can solve to determine the size of the slice of the pie or the angle measure of the circle graph. To do this, we have to remember that there are *360 degrees in a circle* and that the percent of each response will take up the same percentage of our circle, or that percent times 360 degrees. We solve by multiplying 360 by the percent. To solve for the angle measure that represents the number of students who chose pizza, multiply 360×0.20. The result is 72 degrees.

The piece of the pie chart that represents the percent of students who chose tacos is 12 percent of 360 or 0.12×360, which is equal to 43.2 degrees.

Use mental math to determine the number of degrees in our circle graph that represents the percent of students who chose garden salad. Use a calculator to determine the number of degrees for chicken nuggets, hot dogs, and veggie burgers. (You should have answered garden salad, 36 degrees; chicken nuggets, 126 degrees; hot dogs, 64.8 degrees; and veggie burgers, 18 degrees.) When we add all of these values together, we should get a total of 360 degrees, a complete circle.

Try This

Multiple-Choice Questions

Use this data to answer the following questions. When 50 students were asked to name their favorite flavor of ice cream, 22 students chose vanilla, 15 chose chocolate, 8 chose strawberry, and 5 chose cookies and cream.

1. How would you find the percent of students that chose each flavor?

   A. add 2

   B. subtract 2

   C. multiply by 2

   D. divide by 2

2. What percent of students chose vanilla as their favorite flavor?

   A. 10 percent

   B. 16 percent

   C. 30 percent

   D. 44 percent

3. How many degrees of the circle graph represent the number of students who chose chocolate?

   A. 36 degrees

   B. 58 degrees

   C. 108 degrees

   D. 158 degrees

4. What percent of students did *not* choose cookies and cream?

   A. 10 percent

   B. 60 percent

   C. 70 percent

   D. 90 percent

5. How many degrees of the circle graph represent the number of students who chose vanilla and strawberry?

   A. 36 degrees

   B. 100 degrees

   C. 216 degrees

   D. 360 degrees

Open-Ended Question

Construct a table and a circle graph to represent the results of the ice cream survey. Label your circle graph.

The answers appear on pages 236–237.

ADDITIONAL GRAPHS

Statisticians use additional graphs to display data. These include histograms, dot plots, and box plots.

Consider the following numerical data. The frequency table shows the daily temperature in Jersey City throughout the month of July and the number of days that each temperature was reached. The next 3 sections show how this same data can be represented in different graphical ways.

TEMPERATURES IN JERSEY CITY IN JULY

Temperature (Degrees Fahrenheit)	Number of Days
80°	2
81°	4
82°	3
83°	4
84°	5
85°	2
86°	1
87°	0
88°	4
89°	1
90°	2
91°	3

Histogram

This histogram displays the temperature in intervals in relation to the number of days. The temperature intervals are listed horizontally, and the frequency of days is shown vertically.

TEMPERATURES IN JERSEY CITY IN JULY

	80° – 82°	83° – 85°	86° – 88°	89° – 91°
12				
11		▓		
10		▓		
9	▓	▓		
8	▓	▓		
7	▓	▓		
6	▓	▓		▓
5	▓	▓		▓
4	▓	▓	▓	▓
3	▓	▓	▓	▓
2	▓	▓	▓	▓
1	▓	▓	▓	▓
0	▓	▓	▓	▓

Temperatures in Degrees Fahrenheit

You can see from this histogram that 9 days had temperatures between 80 and 82 degrees Fahrenheit, 11 days had temperatures between 83 and 85 degrees Fahrenheit. 5 days had temperatures between 86 and 88 degrees Fahrenheit, and 6 days had temperatures between 89 and 91 degrees Fahrenheit.

Dot Plot

We can also represent the data from the frequency table on a dot plot. A dot plot looks like a number line that contains each of the temperatures. Dots are stacked vertically to represent each day that had a particular temperature.

TEMPERATURES IN JERSEY CITY IN JULY

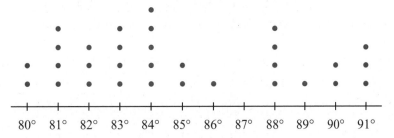

You can use this dot plot to see how many days the temperature was the same.

You have probably seen graphs that look very similar to this that are called line plots. A line plot would use an X instead of a dot to show each time a temperature was reached.

Box Plot

TEMPERATURES IN JERSEY CITY IN JULY

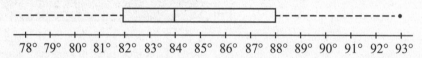

78° 79° 80° 81° 82° 83° 84° 85° 86° 87° 88° 89° 90° 91° 92° 93°

A box plot displays the same data from our temperature frequency table but in a slightly different way. It is designed for you to see the highest and lowest temperatures, the median (middle) value in our data set, and the lower and upper quartile ranges. These ranges show how much of the data is below or above the median.

Let's look at how a box plot is made. First, we have to put all of our data in numerical order. If you look at the frequency table, the temperature reached 80 degrees on two days, so we have to enter 80 two times, enter 81 four times, enter 82 three times, and so on until the temperatures of all thirty-one temperatures are listed for all days. Our data set will look like this:

80 80 81 81 81 81 82 82 82 83 83 83
83 84 84 84 84 84 85 85 86 88 88 88
88 89 90 90 91 91 91

Then we mark the least and greatest values, which are 80 and 91.

Next, we find the median, or the middle number in our data set. That is the 16th number, which is 84. We mark that on the graph.

The next two intervals are the lower and upper quartiles. The lower quartile, which is also called the first quartile, begins at the median point of the lowest number in our data set and ends at the median of all the data. In this case, it is the 8th number in our sequence, or 82 degrees.

The upper quartile begins at the median of all the data and ends at the median of the remaining data entries. In this case, it is the 24th number in our sequence, or 88 degrees.

We box these values to show how they are distributed in quartiles.

Try This

Use the data in this frequency table to construct a histogram, a dot plot, and a box plot.

SIXTH-GRADE MATH FINAL EXAM GRADES

Grade	Students
100	6
95	4
90	5
85	2
80	7
75	4
70	6
65	3
60	2

The answers appear on pages 237–238.

RANGE, MEAN, MEDIAN, AND MODE

When we analyze numerical data, four common values that we work with are range, mean, median, and mode.

Range is the difference between the largest and smallest numbers in our data set. We solve for the range by subtracting the smallest (minimum) number in our set from the largest (maximum) number.

Mean is the average of our set of numbers. We solve for the mean by adding all of the numbers in our data set together and dividing by the number of entries.

Median is the middle number in our data set. We solve for the median by ordering the numbers in our set from least to greatest and selecting the middle number. (When there is an even number of entries in our data set, there is no middle number. In that case, the median is found by averaging the two numbers in the middle of the set.)

Mode is the number that appears most in our data set. Sometimes it is easiest to solve for the mode by ordering our numbers first and seeing which number appears most. Sometimes there is no mode. This will happen when every number appears the same number of times.

For Example

These are Jorge's math quiz scores: 85, 87, 73, 98, 83, 100, 95, 83, 88

What are the range, mean, median, and mode of Jorge's scores?

Range—Subtract the lowest score from the highest. $100 - 73 = 27$.

Mean—Add all scores together and divide by 8. (Use your calculator to do this.)85 + 87 + 73 + 98 + 83 + 100 + 95 + 83 + 88 = 792; 792 ÷ 9 = 88.

Median—Order your numbers first:73, 83, 83, 85, 87, 88, 95, 98, 100.The number 87 is the median because it is the middle number.

Mode—Look at the numbers in order. 83 is the mode because it appears two times while all other numbers appear only once.

Try This

Multiple-Choice Questions

Use this data to answer the questions.

A class in Trenton tracked the temperature in their city for eleven days. The degrees recorded were: 63, 66, 59, 58, 62, 64, 67, 62, 60, 57, 53.

1. What was the temperature range in Trenton during those eleven days?

   A. 11 degrees

   B. 14 degrees

   C. 61 degrees

   D. 62 degrees

2. What was the mean temperature in Trenton for those eleven days?

   A. 11 degrees

   B. 14 degrees

   C. 61 degrees

   D. 62 degrees

3. What was median temperature in Trenton for those eleven days?

   A. 11 degrees

   B. 14 degrees

   C. 61 degrees

   D. 62 degrees

4. What is the mode of the temperatures recorded in Trenton for those eleven days?

   A. 11 degrees

   B. 14 degrees

   C. 61 degree

   D. 62 degrees

Open-Ended Question

The class continued to track the temperature for three more days. The temperature remained at 62 degrees for the three days and only one of the four measures (range, mean, median, and mode) changed. Tell which measure changed and explain why the others remained the same.

The answers appear on page 238.

LESSON 2—PROBABILITY

We use probability to determine how likely it is that an event will occur. When an event is certain to occur, it has a probability of one. When an event is impossible and has no probability of occurring, it has a probability of zero.

Some examples of certain events that have a probability of one are that Tuesday will come after Monday, or that a student who is successful in the sixth grade will be promoted to the seventh grade.

Some examples of events that are impossible and have a probability of zero are rolling a seven on a six-sided number cube, or having a month with 32 days.

Every event that occurs has a probability that is between zero and one. We write this value in the form of a fraction where the numerator is the possibility, or chance of the event occurring, and the denominator is the total number of possibilities. We can also write this value in its decimal form. The more likely an event is to occur, the closer the probability is to one. For example, the probability of a coin landing on heads is *one out of two* possibilities. We say that the probability is $\frac{1}{2}$ or 0.5. If there are 10 markers in a bag and 3 of them are red, the probability that you will select a red marker is 3 out of 10. The probability is $\frac{3}{10}$ or 0.3.

Try This

Multiple-Choice Questions

1. What is the probability of choosing a yellow crayon from a box containing eight different colored crayons?

   A. 0

   B. $\frac{1}{8}$

   C. $\frac{1}{2}$

   D. 1

2. What is the probability that a sixth grader would have a driver's license?

   A. 0

   B. $\dfrac{1}{6}$

   C. $\dfrac{1}{2}$

   D. 1

3. Students must wear sneakers to play in the gym. What is the probability that students in the gym are wearing sneakers?

   A. 0

   B. $\dfrac{1}{18}$

   C. $\dfrac{1}{2}$

   D. 1

4. In a class with an equal number of boys and girls, what is the probability that a girl would be first in line?

   A. 0

   B. $\dfrac{1}{24}$

   C. $\dfrac{1}{2}$

   D. 1

Open-Ended Question

Give one example of an event that has a probability of zero, and one example of an event that has a probability of one.

The answers appear on page 239.

EXPERIMENTAL PROBABILITY

Sometimes data is generated by conducting a series of trials or experiments. Examples of generating experimental data include tossing a coin and keeping a record of the number of heads or tails, or rolling a number cube and keeping track of how often each number appears.

Jerome rolled a six-sided number cube 25 times with the following results or distributions. When we represent this data as **experimental probability**, we use a fraction format, using the number of times a result occurred as the numerator, and the total number of trials as the denominator.

1—6 times   6 out of 25 times   or $\dfrac{6}{25}$

2—4 times   4 out of 25 times   or $\dfrac{4}{25}$

3—5 times   5 out of 25 times   or $\dfrac{5}{25}$ simplified to $\dfrac{1}{5}$

4—3 times   3 out of 25 times   or $\dfrac{3}{25}$

5—2 times   2 out of 25 times   or $\dfrac{2}{25}$

6—5 times   5 out of 25 times   or $\dfrac{5}{25}$ simplified to $\dfrac{1}{5}$

Notice that when we add all of these fractions together, their sum is equal to $\frac{25}{25}$, which is equal to 1.

Try This

Multiple-Choice Questions

1. Wilma tossed a coin 15 times and it landed on heads 9 times. What fraction represents the experimental probability?

   A. $\frac{1}{5}$

   B. $\frac{2}{5}$

   C. $\frac{3}{5}$

   D. $\frac{4}{5}$

2. In Wilma's experiment, what fraction represents the experimental probability of the coin landing on tails?

   A. $\frac{1}{5}$

   B. $\frac{2}{5}$

   C. $\frac{3}{5}$

   D. $\frac{4}{5}$

3. Study the data from Jerome's number-cube-rolling experiment above. What fraction represents the probability that he did *not* roll a 6?

A. $\dfrac{1}{5}$

B. $\dfrac{2}{5}$

C. $\dfrac{3}{5}$

D. $\dfrac{4}{5}$

4. Jerome continued his experiment from the example above. He rolled the number cube an additional five times, and each time rolled a 4. What fraction represents the number of times 4 appeared?

A. $\dfrac{4}{25}$

B. $\dfrac{7}{25}$

C. $\dfrac{4}{30}$

D. $\dfrac{8}{30}$

5. Using the data from Jerome's new experiment, what fraction represents the times he rolled a 1?

A. $\dfrac{1}{5}$

B. $\dfrac{7}{30}$

C. $\dfrac{3}{10}$

D. $\dfrac{1}{2}$

The answers appear on page 239.

RATIOS AND PROPORTIONAL RELATIONSHIPS

LESSON 1—RATIOS AND PROPORTIONS

A ratio represents the relationship between the values of two sets. For example, if there are 12 boys and 15 girls in your class, the ratio of boys to girls is 12 to 15. We can write this as 12:15, or as the fraction $\frac{12}{15}$, or in its simplest form, $\frac{4}{5}$. If we used this same example, we could say that the ratio of boys to the total number of students is 12 to 27. We can also express this as 12:27, $\frac{12}{27}$, or $\frac{4}{9}$. The ratio of girls to the total number of students is 15 to 27, or 15:27, or $\frac{15}{27}$, or $\frac{5}{9}$.

Think About This

Represent each of these numerical relationships as ratios in three ways.

1. The number of inches in a foot to the number of inches in a yard

2. The number of months in a year to the number of seconds in a minute

3. The number of sides in a triangle to the number of seasons in a year

The answers appear on page 240.

Proportions

A **proportion** is a comparison of two ratios. If the ratio of boys to girls in your class is 1 to 2, it means that there is one boy to every two girls. This means that if there were two boys there would be four girls. We multiplied both numbers—the numerator and the denominator—by two. To extend this pattern, the ratio of boys to girls could be 3 to 6, 4 to 8, 5 to 10, and so on. These are all equal ratios.

Think About This

1. The ratio of the red cards to black cards in a deck is 1 to 1. How many black cards are there if there are 26 red cards? Solve by setting up the proportion

$$\frac{1 \text{ red card}}{1 \text{ black card}} = \frac{26 \text{ red cards}}{? \text{ black cards}}$$

2. If one candy bar costs 75¢, how many candy bars can you buy with $6.00? Solve by setting up the proportion

$$\frac{1 \text{ candy bar}}{75¢} = \frac{? \text{ candy bar}}{\$6.00}$$

3. Sandy can read twelve pages in thirty minutes. How long will it take her to read forty-eight pages? Solve by setting up the proportion

$$\frac{12 \text{ pages}}{30 \text{ minutes}} = \frac{48 \text{ pages}}{? \text{ minutes}}$$

4. Ana's mother made forty-eight tacos for the class's international dinner. If twenty-four people attended, how many tacos did each person eat? Solve by setting up the proportion

$$\frac{48 \text{ tacos}}{24 \text{ people}} = \frac{? \text{ tacos}}{1 \text{ person}}$$

The answers appear on page 240.

Try This

Multiple-Choice Questions

1. What is the ratio of the number of vowels to the total number of letters in the alphabet?

   A. 1:26

   B. 5:26

   C. 5:21

   D. 1:25

2. What is the ratio of the number of hours in a day to the number of days in a week?

   A. 12:5

   B. 24:5

   C. 7:24

   D. 24:7

3. Which choice is *not* the ratio of the number of states in the United States to the number of pennies in a dollar?

A. $\dfrac{1}{2}$

B. 50:100

C. 1:1

D. 2 to 4

4. The ratio of dogs to cats in the pet store is 3 to 2. How many cats are there if there are 12 dogs in the store?

A. 2

B. 6

C. 8

D. 12

5. The ratio of quarters to nickels in Willy's piggy bank is 4 to 5. How many quarters does Willy have if he has 25 nickels?

A. 5

B. 10

C. 20

D. 25

6. The ratio of the length of the sides in a rectangle is 4 inches to 6 inches. If the first side increases to 12 inches, what will be the length of the second side?

    A. 3 inches

    B. 12 inches

    C. 18 inches

    D. 24 inches

7. If you can buy six cans of soda for $1.50, how much will eighteen cans of soda cost?

    A. $0.25

    B. $3.00

    C. $4.50

    D. $6.00

Open-Ended Question

The ratio of the length to the width of your school picture is 3 inches by 5 inches. You want to display it in a frame that measures one foot by one and one-half feet. Can the picture be enlarged so that it is in proportion to the frame? Draw a picture and explain your answer.

The answers appear on pages 240–241.

LESSON 2—UNIT RATES

A ratio compares the amount of two different substances. For example, suppose that 10 pencils cost $1.50. We can represent this as the ratio $\dfrac{\$1.50}{10}$.

A unit rate is a special type of ratio. It is the amount for 1. So if 10 pencils cost $1.50, the unit rate is $0.15 per pencil. We can represent this unit rate as the ratio $\dfrac{\$0.15}{1}$.

Patricia ran 2 miles in 30 minutes. To determine the unit rate, or the number of miles she ran in 1 hour, think of 30 minutes as $\dfrac{1}{2}$ hour. The ratio is $\dfrac{2}{\frac{1}{2}}$. Simplify to find

the unit rate. The unit rate is $\dfrac{4}{1}$ or 4 miles per hour.

Try This

1. A six-pack of soda costs $2.94. What is the unit price, or the cost of one can of soda?

2. A family pack of amusement park tickets costs $125.00. What is the unit rate per ticket if a family of five go to the park?

The answers appear on page 241.

Chapter 6

SAMPLE TESTS

ANSWER SHEET: SAMPLE TEST 1 (DAY 1)

Part I: Short Constructed-Response Questions

1. _____

2. _____

3. _____

4. _____

5. _____

6. _____

7. _____

8. _____

Part II: Multiple-Choice Questions

1. Ⓐ Ⓑ Ⓒ Ⓓ　　　5. Ⓐ Ⓑ Ⓒ Ⓓ　　　9. Ⓐ Ⓑ Ⓒ Ⓓ

2. Ⓐ Ⓑ Ⓒ Ⓓ　　　6. Ⓐ Ⓑ Ⓒ Ⓓ　　　10. Ⓐ Ⓑ Ⓒ Ⓓ

3. Ⓐ Ⓑ Ⓒ Ⓓ　　　7. Ⓐ Ⓑ Ⓒ Ⓓ　　　11. Ⓐ Ⓑ Ⓒ Ⓓ

4. Ⓐ Ⓑ Ⓒ Ⓓ　　　8. Ⓐ Ⓑ Ⓒ Ⓓ　　　12. Ⓐ Ⓑ Ⓒ Ⓓ

Part III: Extended Constructed-Response Questions

1.

2.

ANSWER SHEET:
SAMPLE TEST 1 (DAY 2)

Part I: Short Constructed-Response Questions

1. _____

2. _____

3. _____

4. _____

5. _____

6. _____

7. _____

8. _____

Part II: Multiple-Choice Questions

1. Ⓐ Ⓑ Ⓒ Ⓓ   5. Ⓐ Ⓑ Ⓒ Ⓓ   9. Ⓐ Ⓑ Ⓒ Ⓓ

2. Ⓐ Ⓑ Ⓒ Ⓓ   6. Ⓐ Ⓑ Ⓒ Ⓓ   10. Ⓐ Ⓑ Ⓒ Ⓓ

3. Ⓐ Ⓑ Ⓒ Ⓓ   7. Ⓐ Ⓑ Ⓒ Ⓓ   11. Ⓐ Ⓑ Ⓒ Ⓓ

4. Ⓐ Ⓑ Ⓒ Ⓓ   8. Ⓐ Ⓑ Ⓒ Ⓓ   12. Ⓐ Ⓑ Ⓒ Ⓓ

Part III: Extended Constructed-Response Questions

1.

2.

SAMPLE TEST 1 (DAY 1)

PART I: SHORT-CONSTRUCTED RESPONSE QUESTIONS

Directions: Solve each problem and write your answer on the answer sheet. Calculators are not allowed for this section.

1. What is the least common multiple (LCM) of 4 and 6?

2. How much change would you receive from a twenty dollar bill if you bought a cap that cost $12.49?

3. What is the area of a rectangle that has a length of 4.5 centimeters and a width of 3.4 centimeters?

4. Solve: $\dfrac{8}{9} \div \dfrac{2}{5}$

5. Winnie ran one-half of a mile in 3 minutes. If she continues at this rate, how long will it take her to run 2 miles?

6. What number comes next in this sequence?

   1, 2, 6, 22, 86, _____

7. Solve: $3 + 2^3 + 7 \times 4$

8. What number added to $7\dfrac{5}{8}$ will result in the sum of $12\dfrac{1}{2}$?

Go On

153

PART II: MULTIPLE-CHOICE QUESTIONS

Directions: Fill in the correct choice on the answer sheet.

1. Six friends want to share a pizza that has been cut into eight slices. They all want to have an equal amount. How much pizza will each person have?

   A. $\frac{6}{8}$ slice

   B. 1 slice

   C. 1 and $\frac{1}{3}$ slice

   D. 2 slices

2. What is the area of this trapezoid?

   A. 52 m²
   B. 80 m²
   C. 144 m²
   D. 320 m²

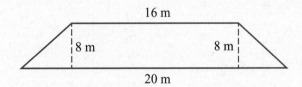

3. Evaluate the expression $4 + n$ when $n = 25$.

   A. 4
   B. 21
   C. 24
   D. 29

4. A dozen donuts costs $9.60. What is the unit rate?

   A. $0.12 per donut

   B. $0.80 per donut

   C. $0.96 per donut

   D. $1.00 per donut

5. Mike's father had 16 magazines in his office. He asked Mike to bring 25 percent of them to the den. How many magazines were left in the office?

   A. 0

   B. 4

   C. 12

   D. 16

6. Kris has 5 books in her backpack and it weighs a total of 30 pounds. If her science book weighs 8 pounds, and her math book weighs 7 pounds, what is the average weight of each of the remaining three books?

   A. 3 pounds

   B. 5 pounds

   C. 6 pounds

   D. 10 pounds

Go On

7. Rami wants to enlarge a picture of his family. The original print has a length of 4 inches and a width of 6 inches. If the length of the enlarged picture is 12 inches, what will the width be?

   A. 6 inches

   B. 12 inches

   C. 18 inches

   D. 24 inches

8. What is the *range* of this set of numbers?
   16, 28, 57, 35, 22, 10, 16.

   A. 16

   B. 22

   C. 26

   D. 47

9. Estimate the product of 288×315.

   A. 900

   B. 9,000

   C. 90,000

   D. 900,000

10. There are 13 boys and 12 girls in Ms. Sani's class. Which statement about her class is true?

    A. Each student represents 4 percent of the class.

    B. There is 1 percent more boys than girls.

    C. The ratio of boys to girls is 13:25.

    D. More than 50 percent of the class is made up of girls.

11. Insert the correct symbol: $3 + 7$ _____ $12 - 5$.

    A. $<$

    B. $>$

    C. $=$

    D. $\approx$

12. The temperature on Monday was 72 degrees, and it was 58 degrees on Tuesday. Which integer represents the change in temperature from Monday to Tuesday?

    A. –14

    B. 14

    C. 58

    D. 72

Go On

PART III: EXTENDED CONSTRUCTED-RESPONSE QUESTIONS

Directions: Show your work on the answer sheet.

1. The frequency table below shows the grades for the students in Mr. Drucker's class. Create a dot plot using this data.

MR. DRUCKER'S CLASS GRADES

100	1
96	3
92	4
88	3
84	1
80	4
76	2
72	6

2. The Pino's have 100 yards of material to fence in an area of their yard as a dog pen. They want their dog, Zac, to be able to run the greatest possible distance. Design and draw a dog pen using the amount of material they have, and label the length and width. Explain your reasoning.

Stop

If you have time, you may review your work in this section only.

SAMPLE TEST 1 (DAY 2)

PART I: SHORT CONSTRUCTED-RESPONSE QUESTIONS

Directions: Solve each problem, and write your answer on the answer sheet. Calculators are not allowed for this section.

1. Three-fifths of the class wants to take a field trip to Liberty Science Center. What decimal represents this amount?

2. Ms. Williams announced that 87 percent of the students in the sixth grade received an A or a B on their math test. What percent of the students did *not*?

3. Solve: $|-3.8|$

4. Solve: $\dfrac{8}{15} \div \dfrac{12}{25}$

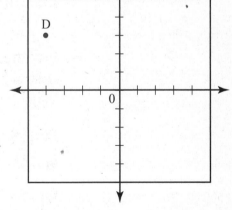

5. Solve: 6.19×14.8

6. Name the coordinates of point *D*.

7. Solve: $2.82 \div 0.12$

8. What is the greatest common factor (GCF) of 10 and 15?

Go On

PART II: MULTIPLE-CHOICE QUESTIONS

Directions: Fill in the correct choice on the answer sheet.

1. The ratio of quarters to nickels in Lee's piggy bank is 4 to 5. How many quarters does Lee have if he has 25 nickels?

   A. 20

   B. 40

   C. 80

   D. 125

2. Jessica is reading a book that has 490 pages. If she wants to finish the book in one week, what is the unit rate of pages she must read per day?

   A. 70

   B. 80

   C. 90

   D. 100

3. Five new video games went on sale last week. They cost $32.25, $38.75, $41.00, $36.15, and $34.50. What is their mean price?

   A. $8.75

   B. $36.15

   C. $36.53

   D. $182.65

4. A total of 4^3 dogs and 3^2 cats were entered in the county pet show. How many dogs and cats were there altogether?

   A. 18

   B. 25

   C. 72

   D. 73

5. The scale on a map is 1 centimeter to 10 kilometers. What is the actual distance between two cities that are 9 centimeters apart on the map?

   A. 9 kilometers

   B. 10 kilometers

   C. 90 kilometers

   D. 9,000 centimeters

6. A rectangle has an area of 75 square centimeters and a length of 6 centimeters. What is its width?

   A. 8 centimeters

   B. 12.5 centimeters

   C. 81 centimeters

   D. 30 centimeters

7. According to the distributive property:
   $8 \times (3.1 + 4.2) =$

   A. 8×13.02

   B. $(8 + 3.1) \times (8 + 4.2)$

   C. $(8 \times 3.1) + (8 \times 4.2)$

   D. $(8 \times 8) + (3.1 \times 4.2)$

   Go On

8. Estimate the quotient: $0.03 \div 1.001 =$

   A. 0.03

   B. 0.30

   C. 3.00

   D. 30.0

9. Kumar bought a sandwich for $3.95, a soda for $1.15, and chips for $0.60. How much change did Kumar get back from $6.00?

   A. 2 quarters

   B. 2 dimes and 5 pennies

   C. 5 nickels and 5 pennies

   D. 1 quarter and 1 dime

10. What numbers complete this table?

n	30	25	20	15	10
$n - 3.5$	26.5				

   A. 21.5, 16.5, 11.5, 6.5

   B. 30.5, 28.5, 23.5, 18.5, 13.5

   C. 23, 19.5, 16, 12.5

   D. 6.5, 3, −0.5, −4

11. Solve: $6\frac{3}{4} - 5\frac{1}{2} =$

   A. $\frac{3}{4}$

   B. 1.25

   C. $1\frac{1}{2}$

   D. 12.25

12. How much wrapping paper would you need to cover a box that has a length of 30 centimeters, a width of 25 centimeters, and a height of 10 centimeters?

   A. 260 square centimeters

   B. 1,400 square centimeters

   C. 2,600 square centimeters

   D. 7,500 square centimeters

Go On

PART II: EXTENDED CONSTRUCTED-RESPONSE QUESTION (1)

Directions: Show your work on the answer sheet.

Paulo surveyed 100 students in his school to determine their favorite colors. The table below shows their responses.

Color	Number of students
Red	18
Green	10
Blue	35
Brown	12
Purple	5
Orange	20

Study the table and answer the following questions.

1. What is the range of the data in this table?

2. How many students out of 100 said that blue was their favorite color?

   A. Express this value as a percent.

   B. Express this value as a decimal.

   C. Express this value as a fraction in lowest terms.

3. Construct a histogram to represent this data. Be sure to label your graph.

PART II: EXTENDED CONSTRUCTED-RESPONSE QUESTION (2)

Directions: Show your work on the answer sheet.

Marcia is getting ready to plant a flower garden. She wants to design one that will provide the greatest area. These are her choices:

A. A circular garden with a radius of 8 feet

B. A rectangular garden with a length of 28 feet and a width of 7 feet

1. Draw a diagram for each of these designs and label the dimensions.

2. Calculate the area of each design. Which garden should Marcia plant? Show your work and label the answers accurately.

3. If the materials needed cost $1.25 per square foot, how much will it cost for Marcia to build the garden?

PART II: EXTENDED CONSTRUCTED-RESPONSE QUESTION (3)

Directions: Show your work on the answer sheet.

Roger is reading a 250-page book.

On the first day, he read 20 pages.

On the second day, he read 5 more pages than he read on the first day.

On the third day, he read 10 more pages than he read on the second day.

On the fourth day, he read 15 more pages than he read on the third day, and so on.

1. Complete this table to indicate the number of pages Roger read each day.

Day	Number of Pages
1	20
2	
3	
4	
5	

2. What is the total number of pages Roger read by the end of the fifth day?

3. If Roger can read 25 pages in an hour, how many hours will he have to read to finish the book on the sixth day?

Stop

If you have time, you may review your work in this section only.

SAMPLE TEST 1—SOLUTIONS (DAY 1)

PART I: SHORT CONSTRUCTED-RESPONSE QUESTIONS

1. The LCM of 4 and 6 is 12.

2. $20.00 − 12.49 = $7.51

3. Apply the formula $a = \ell \times w = (4.5 \text{ cm} \times 3.4 \text{ cm} = 15.3 \text{ cm}^2)$

4. $2\dfrac{2}{9}$

5. Winnie runs at the rate of 1 mile every 6 minutes. 2 miles takes 12 minutes.

6. The rule is multiply by 4 and subtract 2. You get 342.

7. 39

8. $4\dfrac{7}{8}$

PART II: MULTIPLE-CHOICE QUESTIONS

1. **C** 1 and $\dfrac{1}{3}$ slice. 8 slices divided by 6 people is equal to 1 and $\dfrac{1}{3}$.

2. **C** 144 m^2

3. **D** 29. 4 + 25 = 29.

4. **B** $0.80 per donut.

5. **C** 12. 25 percent of 16 equals 4. If Mike brought 4 magazines, there were 12 left.

6. **B** 5 pounds. The science and math books weigh a total of 15 pounds and the remaining three books weigh a total of 15 pounds. The average weight of these three books is $15 \div 3 = 5$.

7. **C** 18 inches. If the length was increased to three times its size, the width must be also. $6 \times 3 = 18$.

8. **D** 47. Range is the difference between the largest and smallest numbers. $57 - 10 = 47$.

9. **C** 90,000. Round each to 300 and multiply $300 \times 300 = 90,000$.

10. **A** Each student represents 4 percent of the class. $\frac{1}{25} = \frac{4}{100} = 4$ percent.

11. **B** >. Solve each side of the inequality to determine that $10 > 7$.

12. **A** −14. The negative symbol shows a decrease in temperature.

PART III: EXTENDED CONSTRUCTED-RESPONSE QUESTIONS

1.

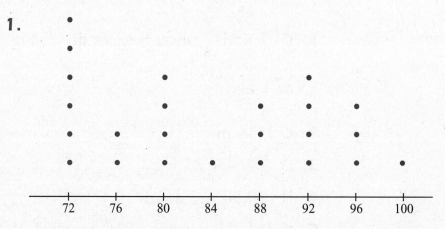

2. The dimensions should be 48 yards × 1 yard. The explanation should indicate that the greatest length would provide the longest distance to run.

SAMPLE TEST 1—SOLUTIONS (DAY 2)

PART I: SHORT CONSTRUCTED-RESPONSE QUESTIONS

1. 0.6

2. 13 percent

3. 3.8

4. $1\frac{1}{9}$

5. 91.612

6. (−4, 3)

7. 23.5

8. 5

PART II: MULTIPLE-CHOICE QUESTIONS

1. A 20. If the proportion of quarters to nickels is 4 to 5 and there are 25 or 5 × 5 quarters, we find the number of nickels by multiplying 4 × 5.

2. A 70. There are seven days in a week; the unit rate is 490 ÷ 70 or 70 pages per day.

3. C $36.53. The mean is the average of all the prices.

4. D 73. $4^3 = 64$ and $3^2 = 9$; 64 + 9 = 73

5. C 90 kilometers. The ratio is 1 centimeter to 10 kilometers. Since the number of centimeters was multiplied by 9, the number of kilometers is multiplied by 9 also; 10 × 9 = 90 kilometers.

6. **B** 12.5 centimeters. Area = length × width. If l × w = 75, then 6 × w = 75 and 6 × 12.5 = 75. Therefore, the width = 12.5 centimeters.

7. **C** (8 × 3.1) + (8 × 4.2). According to the distributive property, you get the same result when you multiply each term and later add them, as you do when you add them first and then multiply by a common factor.

8. **A** 0.03. If we round the divisor to 1 and divide, we know that any number divided by 1 is equal to itself.

9. **C** 5 nickels and 5 pennies. Kumar's meal cost $5.70 so he should receive 30¢ change from $6.00. 5 nickels and 5 pennies total 30¢.

10. **A** 21.5, 16.5, 11.5, 6.5. Subtracting 3.5 from each number in the first row gives the results in the second row.

11. **B** 1.25. When we subtract $6\frac{3}{4} - 5\frac{1}{2}$, the result, or difference, is $1\frac{1}{4}$, which is equal to 1.25 as a decimal.

12. **C** 2,600 square centimeters is the surface area of the box. You solve by adding the areas of all of the surfaces together:

$$top = 30 \times 25 = 750$$
$$bottom = 30 \times 25 = 750$$
$$front = 30 \times 10 = 300$$
$$back = 30 \times 10 = 300$$
$$left\ side = 25 \times 10 = 250$$
$$right\ side = 25 \times 10 = 250$$

Total surface area = 2,600 square centimeters

PART II: EXTENDED CONSTRUCTED-RESPONSE QUESTION (1)

1. The range is 30. Solve this by subtracting the smallest number, 5, from the largest number, 35; $35 - 5 = 30$.

2. 35 students said that blue was their favorite color.

   A) 35 percent   B) 0.35   C) $\dfrac{7}{20}$

3.

	Red	Green	Blue	Brown	Purple	Orange
35			▓			
34			▓			
33			▓			
32			▓			
31			▓			
30			▓			
29			▓			
28			▓			
27			▓			
26			▓			
25			▓			
24			▓			
23			▓			
22			▓			
21			▓			
20			▓			▓
19			▓			▓
18	▓		▓			▓
17	▓		▓			▓
16	▓		▓			▓
15	▓		▓			▓
14	▓		▓			▓
13	▓		▓			▓
12	▓		▓	▓		▓
11	▓		▓	▓		▓
10	▓	▓	▓	▓		▓
9	▓	▓	▓	▓		▓
8	▓	▓	▓	▓		▓
7	▓	▓	▓	▓		▓
6	▓	▓	▓	▓		▓
5	▓	▓	▓	▓	▓	▓
4	▓	▓	▓	▓	▓	▓
3	▓	▓	▓	▓	▓	▓
2	▓	▓	▓	▓	▓	▓
1	▓	▓	▓	▓	▓	▓
0						

PART II: EXTENDED CONSTRUCTED-RESPONSE QUESTION (2)

1.

28 feet

7 feet

8 feet

2. Circle—

$A = \pi r^2$

$A = 3.14 \times 8 \times 8$

$A = 3.14 \times 64$

$A = 200.96$ square feet

Rectangle—

$A = l \times w$

$A = 28 \times 7$

$A = 196$ square feet

Marcia should plant the circular garden because it has a larger area than the rectangular garden.

3. The cost of the circular garden is 200.96 square feet times $1.25.

$200.96 \times 1.25 = 251.2$

The cost of the garden is $251.20.

PART II: EXTENDED CONSTRUCTED-RESPONSE QUESTION (3)

1.

Day	Number of Pages
1	20
2	25
3	35
4	50
5	70

2. Roger read 200 pages by the end of the fifth day.

3. 2 hours. Roger has to read an additional 50 pages to finish the book on the sixth day. This will take two hours at the rate of 25 pages per hour.

ANSWER SHEET:
SAMPLE TEST 2 (DAY 1)

Part I: Short Constructed-Response Questions

1. _____

2. _____

3. _____

4. _____

5. _____

6. _____

7. _____

8. _____

Part II: Multiple-Choice Questions

1. Ⓐ Ⓑ Ⓒ Ⓓ   5. Ⓐ Ⓑ Ⓒ Ⓓ   9. Ⓐ Ⓑ Ⓒ Ⓓ

2. Ⓐ Ⓑ Ⓒ Ⓓ   6. Ⓐ Ⓑ Ⓒ Ⓓ   10. Ⓐ Ⓑ Ⓒ Ⓓ

3. Ⓐ Ⓑ Ⓒ Ⓓ   7. Ⓐ Ⓑ Ⓒ Ⓓ   11. Ⓐ Ⓑ Ⓒ Ⓓ

4. Ⓐ Ⓑ Ⓒ Ⓓ   8. Ⓐ Ⓑ Ⓒ Ⓓ   12. Ⓐ Ⓑ Ⓒ Ⓓ

Part III: Extended Constructed-Response Questions

1.

2.

ANSWER SHEET: SAMPLE TEST 2 (DAY 2)

Part I: Short Constructed-Response Questions

1. _____

2. _____

3. _____

4. _____

5. _____

6. _____

7. _____

8. _____

Part II: Multiple-Choice Questions

1. Ⓐ Ⓑ Ⓒ Ⓓ   5. Ⓐ Ⓑ Ⓒ Ⓓ   9. Ⓐ Ⓑ Ⓒ Ⓓ

2. Ⓐ Ⓑ Ⓒ Ⓓ   6. Ⓐ Ⓑ Ⓒ Ⓓ   10. Ⓐ Ⓑ Ⓒ Ⓓ

3. Ⓐ Ⓑ Ⓒ Ⓓ   7. Ⓐ Ⓑ Ⓒ Ⓓ   11. Ⓐ Ⓑ Ⓒ Ⓓ

4. Ⓐ Ⓑ Ⓒ Ⓓ   8. Ⓐ Ⓑ Ⓒ Ⓓ   12. Ⓐ Ⓑ Ⓒ Ⓓ

Part III: Extended Constructed-Response Questions

1.

2.

3.

SAMPLE TEST 2 (DAY 1)

PART I: SHORT-CONSTRUCTED RESPONSE QUESTIONS

Directions: Solve each problem and write your answer on the answer sheet. Calculators are not allowed for this section.

1. What is the greatest common factor (GCF) of 21 and 36?

2. Solve: $6\dfrac{2}{5} \times 4\dfrac{3}{8}$

3. Use the order of operations to solve:
$3 + 4^2 \div 2 - 5 \times 2 =$

4. What is the surface area of a cube with a side length of 2.75 cm?

5. Solve: $3\dfrac{3}{8} - 2\dfrac{1}{4}$

6. What number comes next in this sequence?
8, 4, 0, –4, _____

7.

What rational number is at point A?

8. Solve 896 ÷ 20. Represent the quotient with the remainder as a whole number, a fraction, and a decimal.

Go On

PART II: MULTIPLE-CHOICE QUESTIONS

Directions: Fill in the correct answer on the answer sheet.

1. Which of these values is closest to zero?

   A. $\dfrac{6}{8}$

   B. 1

   C. −1

   D. $-\dfrac{1}{3}$

2. What is the difference of $1 - \dfrac{3}{5}$

   A. $\dfrac{1}{2}$

   B. $\dfrac{2}{3}$

   C. $\dfrac{2}{5}$

   D. $\dfrac{3}{5}$

3. Evaluate the expression $4 \times n + 6$ when $n = 6$.

   A. 4

   B. 21

   C. 24

   D. 30

4. Which of these numbers is less than $-\dfrac{1}{2}$?

   A. 2

   B. 1

   C. 0

   D. −1

5. There are 28 students in Ms. Valere's class. 50 percent of them received an A on their exam. How many students got an A?

   A. 10

   B. 14

   C. 18

   D. 26

6. What is the difference of $96 - 4.879$?

   A. 91.121

   B. 47.21

   C. 4.721

   D. 4783

Go On

7. Rahul wants to serve empanadas to his friends at 5:30. It takes 30 minutes to make the filling, and 20 minutes for them to cook. At what time should he begin preparing them?

A. 4:20

B. 4:40

C. 5:00

D. 5:10

8. What is the *median* of this set of numbers? 16, 28, 57, 35, 22, 10, 16.

A. 16

B. 22

C. 26

D. 47

9. Estimate the sum of 288.8 + 3.156.

A. 30

B. 300

C. 500

D. 3000

10. What is the sum of 364.932 + 25.07?

A. 367.439

B. 390.002

C. 3.90002

D. 251064.93

11. What is the value of the number 8 in 235.098435?

    A. 8 thousand

    B. 8

    C. 8 hundredths

    D. 8 thousandths

12. Joshua entered the elevator on the first floor. It went up 3 levels, down 2, up 9, and down 8 floors before he got out. Which set of integers represents his elevator ride?

    A. +3, +2, +9, +8

    B. −3, −2, −9, −8

    C. +3, −2, +9, −8

    D. −3, +2, −9, +8

Go On

PART III: EXTENDED CONSTRUCTED-RESPONSE QUESTIONS

Directions: Show your work on the answer sheet.

1. Rachel wants to earn $75.00 so she can buy a season ticket to Adventureland Amusement Park. Her parents agree to pay her $2.00 each time she dries the dishes, $3.00 for each hour she babysits her younger brother, and $1.00 every time she walks the dog. Make a plan for Rachel so that she can earn the $75.00 in less than three weeks. Show your plan in a chart and explain it.

2. Sanjaya has a piece of wrapping paper that measures 1,440 square inches. He is using it to wrap a present that is in a rectangular box that is 10 inches long, 20 inches wide, and 6 inches deep. How many square inches of wrapping paper will he have left? Draw a diagram of the box and show your work.

Stop

If you have time, you may review your work in this section only.

SAMPLE TEST 2 (DAY 2)

PART I: SHORT CONSTRUCTED-RESPONSE QUESTIONS

Directions: Solve each problem, and write your answers on the answer sheet. Calculators are not allowed in this section.

1. What is the unit rate of an airplane flying 1,200 miles in 6 hours?

2. Water freezes at 0 degrees Celsius. The temperature of the water in Mark's experiment is 4 degrees Celsius. Which integer represents the change in temperature needed for Mark's water to freeze?

3. Solve: $10.26 \div 0.27$

4. Two sides of a triangle measure 3.5 centimeters and 7.2 centimeters. What is the measure of the third side if the perimeter is 16 centimeters ?

5. What is the mean of this set of data: 9.2, 9.4, 10.6, 8.7, 7.9, 8.2?

6. Order these fractions from least to greatest.
   $$\frac{1}{19}, \frac{8}{9}, \frac{19}{23}, \frac{9}{16}$$

7. Julio is building a dog pen that requires 18 feet of fencing. He already has 13 feet 8 inches of fencing. How much more does Julio need?

8. RB Music Center is having a sale. It will take 20 percent off the price of all CDs regularly priced above $15.00. How much will Suzie pay for a CD that originally cost $22.00?

Go On

PART II: MULTIPLE-CHOICE QUESTIONS

1. What two values are equivalent to 24 percent?

   A. 0.24 and $\dfrac{6}{25}$

   B. 0.25 and $\dfrac{24}{100}$

   C. 0.625 and $\dfrac{24}{100}$

   D. 2.4 and $\dfrac{6}{25}$

2. Evaluate the expression $5 \times n + 17$ when n is equal to 4.

   A. 22
   B. 37
   C. 54
   D. 71

3. What is the volume of a rectangular prism when l = 3.5 centimeters, w = 2.4 centimeters, and h = 1.2 centimeters?

   A. 7.1 cubic centimeters
   B. 9.95 cubic centimeters
   C. 10.08 cubic centimeters
   D. 30.96 cubic centimeters

4. Evaluate the expression $18 - 4^2 + 8$.

   A. 10

   B. 22

   C. 34

   D. 204

5. How much change should you receive from a 100-dollar bill if you buy a video game that costs $56.00 and there is 7 percent tax?

   A. $40.08

   B. $43.08

   C. $44.00

   D. $59.92

6. At last year's field day, fourteen students each ran 106.08 meters in a relay race. How far from 1,500 meters did they run?

   A. 4.492 meters

   B. 14 meters and 88 centimeters

   C. 148 meters and 8 centimeters

   D. 1.485 kilometers

7. How many buses are needed for 175 sixth graders to go on a class trip if each bus holds 45 students?

   A. 3.88

   B. 3 r40

   C. $3\dfrac{8}{9}$

   D. 4

Go On

8. Which fraction should you multiply the fraction $\frac{7}{8}$ by to get a product of 1?

   A. $\frac{1}{1}$

   B. $\frac{1}{8}$

   C. $\frac{7}{8}$

   D. $\frac{8}{7}$

9. Evaluate the expression $x + \frac{1}{4}$ when $x = \frac{1}{4}$.

   A. 0

   B. $\frac{1}{2}$

   C. $\frac{1}{8}$

   D. $\frac{2}{8}$

10. Caroline's bank statement shows a balance of −$5.00. How much does Caroline have to deposit so she can write a check for $15.00?

    A. −$5.00

    B. $20.00

    C. $25.00

    D. $35.00

11. Solve: $24 + 30 \div 5 - 8 \times 2$.

    A. 5.6

    B. 14

    C. 44

    D. 75

12. The graph below represents the number of students who attended Homework Club on each of five days. About how many more students attended on day 2 than on day 5?

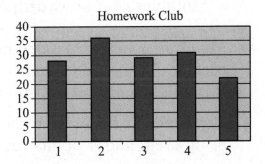

    A. 14

    B. 22

    C. 34

    D. 46

PART III: EXTENDED CONSTRUCTED-RESPONSE QUESTION (1)

Directions: Show your work on the answer sheet.

Mike and Carl are playing a game on this game board. The players toss a marker onto the board and receive points according to these rules.

- Even numbers are worth 1 point.
- Odd numbers are worth 2 points.
- Prime numbers are worth 3 points.
- Multiples of 3 are worth 4 points.
- Multiples of 5 are worth 5 points.

Mike's first marker landed on 7. He received 2 points because 7 is an odd number, and 3 more points because it is prime, for a total of 5 points.

Carl landed on 12. He received 5 points, 1 point for being even, and 4 points for being a multiple of 3.

22	12	5	20	3
14	1	24	15	9
6	19	8	23	17
21	4	16	11	13
10	18	2	25	7

Use the rules of the game to answer the following questions.

1. How many points will Mike receive if he lands on the number 2? Explain your answer.

2. How can a player score exactly 1 point? Explain your reasoning.

3. Carl needs 11 points to win. What number should he land on and why?

Go On

PART III: EXTENDED CONSTRUCTED-RESPONSE QUESTION (2)

Directions: Show your work on the answer sheet.

Use this set of numbers to answer the following questions.

8, 23, 29, 21, 19, 30, 26, 30, 30

1. Which is greater—the range or the median of this set?

2. Calculate the mean of this set of numbers. Show your work.

3. What additional number can be included in this set that will result in a mean that is equal to 24? Show your work and explain your answer.

PART III: EXTENDED CONSTRUCTED-RESPONSE QUESTION (3)

Directions: Show your work on the answer sheet.

Rafael was shopping for sweatshirts that cost $20.00 each.

1. How many sweatshirts can Rafael buy with $150.00? Show your work.

While Rafael was shopping, a special sale was announced. For twenty minutes, the price of everything in the store was reduced by 20 percent. He wanted to buy a pair of jeans that were originally priced at $40.00.

2. How much did the jeans cost during the sale? Show your work.

3. After the twenty-minute special sale, the prices were adjusted by increasing the sale price by 20 percent. What was the price of the jeans then? Explain your answer.

Stop

If you have time, you may review your work in this section only.

SAMPLE TEST 2—SOLUTIONS (DAY 1)

PART I: SHORT CONSTRUCTED-RESPONSE QUESTIONS

1. The GCF of 21 and 36 is 3.

2. 28

3. $3 + (16 \div 2) - (5 \times 2) = 3 + 8 - 10 = 1$.

4. 45.375 cm². The area of one face of the cube is 2.75 × 2.75 cm or 7.5625 cm². A cube has 6 faces. The surface area 7.5625 × 6 = 45.375 cm².

5. $1\dfrac{1}{8}$

6. −8. The rule is to subtract 4 from the previous number.

7. $-\dfrac{1}{2}$

8. 44r 16, $44\dfrac{4}{5}$, 44.8.

PART II: MULTIPLE-CHOICE QUESTIONS

1. D $-\dfrac{1}{3}$

2. C $\dfrac{2}{5}$

3. D 30. 4 × 6 = 24 and 24 + 6 = 30.

4. D −1

5. **B** 14. 50 percent of 28 = 14.

6. **A** 91.121

7. **B** 4:40. The total prep time is 50 minutes, and the time elapsed between 4:40 and 5:30 is 50 minutes.

8. **B** 22. The median is the middle number of the set when they are ordered from least to greatest.

9. **B** 300.

10. **B** 390.002

11. **D** 8 thousandths.

12. **C** +3, −2, +9, −8

PART III: EXTENDED CONSTRUCTED-RESPONSE QUESTIONS

1. Answers will vary. All solutions should result in a total of $75.00. The amounts should be included, calculated, and displayed in a table similar to this one:

Dishes	Babysitting/Hours	Dog Walking
10 @ 2.00	15 hours @ 3.00	10 @ 1.00

2. The surface area of the box is 760 square inches. He will have 680 square inches left.

$$\text{Surface Area} = 2(10 \times 20) = 400$$
$$2(20 \times 6) = 240$$
$$2(10 \times 6) = 120$$
$$\text{Total} = 760 \text{ square inches}$$
$$1440 - 760 = 680 \text{ square inches}$$

SAMPLE TEST 2 SOLUTIONS (DAY 2)

PART I: SHORT CONSTRUCTED-RESPONSE QUESTIONS

1. 200 miles per hour

2. −4 degrees Celsius

3. 38

4. 5.3 centimeters. The sum of the three sides must equal the perimeter of 16 centimeters. $3.5 + 7.2 = 10.7$. Adding 5.3 to 10.7 gives 16.

5. 9.00. The mean is the average of the set of numbers. Solve by adding them all together and dividing the sum by 6.

6. $\frac{1}{19}, \frac{9}{16}, \frac{19}{23}, \frac{8}{9}$. Estimate the value of each of the fractions—$\frac{1}{19} \approx 0$, $\frac{9}{16} \approx \frac{1}{2}$, $\frac{19}{23} \approx \frac{3}{4}$, and $\frac{8}{9} \approx 1$. Order the whole number equivalents from least to greatest.

7. 4 feet 4 inches. Subtract 13 feet 8 inches from 18 feet. Regroup 18 feet as 17 feet 12 inches and subtract the feet from feet and the inches from inches.

8. $17.60. Solve for 20 percent of $22.00 by multiplying $22 \times .20$. Subtract that result, $4.40, from $22.00 to get $17.60.

PART II: MULTIPLE CHOICE QUESTIONS

1. A 0.24 and $\frac{6}{25}$. Percent means hundredth. 24 hundredths written in decimal form is 0.24 and as a fraction is $\frac{24}{100}$. When we simplify this fraction by

dividing the numerator and denominator by their greatest common factor of 4, the result is $\frac{6}{25}$.

2. **B** 37. When we substitute 4 for *n* and solve, $5 \times 4 = 20 + 17 = 37$.

3. **C** 10.08 cubic centimeters. Solve for the volume of a rectangular prism by applying the formula $V = l \times w \times h$. Substitute the values into the formula $V = 3.5 \times 2.4 \times 1.2$ and solve. The solution is 10.08, and it is labeled in cubic centimeters.

4. **A** 10. Solve by applying the rules for the order of operations. First raise 4 to the second power by multiplying 4×4. Subtract that result, 16, from 18 for a difference of 2, and add 8 to get 10.

5. **A** $40.08. First find the tax on $56.00 by multiplying $56 \times .07$. That result is $3.92. Add the tax to the cost of the game, $56.00 + 3.92$, for a total of 59.92. Subtract this amount from $100.00 to get $40.08.

6. **B** 14 meters and 88 centimeters. The total that the students ran was 14 times 106.08 meters or 1,485.12 meters. When you subtract 1,485.12 from 1,500, the difference is 14.88, which is equivalent to 14 meters and 88 centimeters.

7. **D** 4. When you divide the total number of students (175) by the number of students per bus (45) the result is 3 r40. You would need 4 buses, as there cannot be a remainder of a bus.

8. **D** $\frac{8}{7}$. You multiply a fraction by its reciprocal to get the product of 1.

9. B $\frac{1}{2}$. Substitute $\frac{1}{4}$ for x and add $\frac{1}{4} + \frac{1}{4}$ for a sum of $\frac{1}{2}$.

10. B $20.00. Caroline has to deposit $5.00 to bring the balance to zero, and then add an additional $15.00 for the check she wants to write. These two amounts together, $5.00 + $15.00, are equal to $20.00.

11. B 14. Use the order of operations to solve:
$$24 + 30 \div 5 - 8 \times 2$$
$$24 + 6 - 16$$
$$30 - 16$$
$$14$$

12. A 14. According to the graph, about 36 students attended Homework Club on Tuesday, and about 22 attended on Friday. The difference of $36 - 22 = 14$.

PART III: EXTENDED CONSTRUCTED-RESPONSE QUESTION (1)

1. Mike will receive 4 points; 2 is an even number and a prime number.

2. A player can score 1 point by landing on any even number other than 2 that is not a multiple of 3 or 5.

3. Carl should land on 15. He would receive 2 points because it is odd, 4 points because it is a multiple of 3, and 5 points because it is a multiple of 5.

PART III: EXTENDED CONSTRUCTED-RESPONSE QUESTION (2)

1. The range is 22 and the median is 26. The median is greater.

2. 8 + 23 + 29 + 21 + 19 + 30 + 26 + 30 + 30 = 216; 216 ÷ 9 = 24.

3. 24—If we want the mean of 10 numbers to equal 24, the sum of those numbers must equal 240. Since the sum of the original 9 is 216, adding 24 to the set will result in 240.

PART III: EXTENDED CONSTRUCTED-RESPONSE QUESTION (3)

1. Rafael can buy 7 sweatshirts.

2. $32.00. Solve this by finding 20 percent of $40.00 and subtracting that amount from the original price. .20 × 40 = 8; 40 − 8 = 32.

3. $38.40—The new price was found by taking 20 percent of $32.00 and adding it to that price. 20 percent of 32.00 = $6.40; $32.00 + $6.40 = $38.40.

APPENDICES

ANSWER EXPLANATIONS
TO CHAPTER EXERCISES

CHAPTER 1—THE NUMBER SYSTEM

LESSON 1—WHOLE NUMBER PLACE VALUE AND COMPUTATION

Multiple-Choice Questions (page 2)

1. C 600

2. A 1,000,000

3. C 80,000

4. D 4

Short Constructed-Response Question (page 3)

987,541,362

Comparing and Ordering Whole Numbers

Think About This (page 3)

1. 6,421 > 6,412 and 6,412 < 6,421

2. 12,589 < 32,654 and 32,654 > 12,589

3. 15,699 < 15,799 and 15,799 > 15,699

4. 15, 29, 46, 63, 76, 85
    85, 76, 63, 46, 29, 15

202 ■ New Jersey ASK 6 Math Test

5. 29, 209, 290, 902, 920
   920, 902, 290, 209, 29

6. 1,589, 1,598, 1,745, 2,365, 2,662
   2,662, 2,365, 1,745, 1,598, 1,589

Rounding Whole Numbers

Think About This (page 4)

1. 30, 40, and 50

2. 100, 200, 300, 400, and 500;
   1,000, 2,000, 3,000, 4,000, and 5,000

3. 100 and 200/100;
   600 and 700/600

4. 3,000 and 4,000/3,000;
   6,000 and 7,000/7,000

5. 30,000 and 40,000/30,000;
   70,000 and 80,000/80,000

Rounding Decimals

Think About This (page 5)

1. 0   4. 4

2. 1   5. 5

3. $\dfrac{1}{2}$ 6. 136

Decimals as Money

Think About This (page 6)

1. Fifty cents or $0.50

2. $4.00

3. $90.00 (Remember consecutive multiples)

4. $2,000.00

5. $30,000

6. $100,000

7. Answers will vary.

Estimation in Computation

Think About This (page 7)

Round 18 to 20 and $1.09 to $1.00 and estimate the answer of about $20.00. Use a calculator to multiply 18 × 1.09 = $19.62, and compare. Because the exact answer and the estimate are about the same, this is a reasonable amount.

Multiple-Choice Questions (page 7)

1. **B** 400 and 500

2. **C** 23,800

3. **D** proper fractions

4. **C** $3\frac{1}{2}$

5. **A** 27

6. **C** $100.00

Open-Ended Questions (page 9)

1. $1,100.00

2. $5.00

3. division

4. about 220

5. 204.76

6. Letters should explain the process and state that the estimate was unreasonable.

Numerical Operations with Whole Numbers

Think About This (page 10)

1. Multiplication; $232.00

2. Division; 16

3. Subtraction; 5

Whole Number Computation

Try This (page 11)

1. 9,660

2. 1,235,978

3. 61,536,488

4. 4,578,210

5. 2,083,557

6. 11,357,424

Try This (page 12)

1. 35 r10; 35.4; $35\frac{2}{5}$

2. 4 r18 4.25; $4\frac{1}{4}$

3. 297

4. 1,600

5. 91

6. 1146 r13 1146.5; $1146\frac{1}{2}$

Squares and Cubes

Think About This (page 13)

1. $2 \times 2 \times 2 = 8$;
   base 2;
   exponent 3

2. $15 \times 15 = 225$;
   base 15;
   exponent 2

Order of Operations

Think About This (page 15)

1. $10 - 2 \times 6 \div 4 + 5$

   $10 - 12 \div 4 + 5$

   $10 - 3 + 5$

   $7 + 5$

   12

2. $6 \times (2 + 10) \div 3$

   $6 \times 12 \div 3$

   $72 \div 3$

   24

3. $3 \times (2 + 4) \div 3 = 6$

4. $(4 + 1) + 2 \times 5 = 15$ or $4 + 1 + 2 \times 5 = 15$

5. $4 + (1 + 2) \times 5 = 19$

6. $11.6666 \ldots$ The calculator applied the order of operations.

Multiple-Choice Questions

1. A addition

2. C multiplication

3. D 2 r4

4. B exponent

5. D 8

6. B 10

7. B subtract 9 − 2

Extended Constructed-Response Question

The correct answer is 20 r8. The calculator carried out the division to the next decimal place resulting in the quotient 20.5. Ellen did not recognize the decimal as .5 tenths or one-half of the divisor, 16.

LESSON 2—FRACTION PLACE VALUE AND COMPUTATION

Rounding Fractions

Think About This (page 20)

1. 0

2. $\dfrac{1}{2}$

3. 1

4. Answers will vary.

5. 6

6. 2

7. 7

Multiple-Choice Questions (page 21)

1. C numerator

2. B denominator

3. A $\dfrac{1}{4}$

4. C $\dfrac{3}{10}$ and $\dfrac{1}{2}$

5. A $\dfrac{5}{6}$

6. C $3 \div 12$

7. A 0.25

Extended-Constructed Response Question (page 23)

Number line should show equal intervals and include the fractions $0, \dfrac{1}{10}, \dfrac{1}{5}, \dfrac{3}{10}, \dfrac{2}{5}, \dfrac{1}{2}, \dfrac{3}{5}, \dfrac{7}{10}, \dfrac{4}{5}, \dfrac{9}{10}, 1$, and the decimals $0, .1, .2, .3, .4, .5, .6, .7, .8, .9, 1.0$.

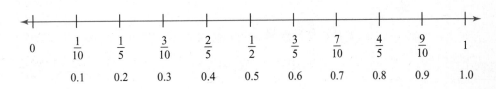

Numerical Operations with Fractions

Factors (page 23)

1. 4 and 8

2. 2 and 0

Think About This (page 24)

3. 1 and 61

4. 1, 2, 3, 6, 7, 14, 21, and 42

5. 1, 5, 25, 125

Think About This—GCF (page 24)

1. 16—1, 2, 4, 8, and 16

36—1, 2, 3, 4, 6, 9, 12, 18, and 36

GCF = 4

2. 28—1, 2, 4, 7, 14, 28

84—1, 2, 3, 4, 7, 12, 21, 28, 42, 84

GCF = 28

3. 65—1, 5, 13, 65

100—1, 2, 4, 5, 10, 20, 25, 50, 100

GCF = 5

Think About This—Multiples (page 25)

1. 7, 14, 21, 28, 35

2. 10, 20, 30, 40, 50

3. 13, 26, 39, 52, 65

Think About This—LCM (page 25)

1. 45
2. 24
3. 36

Multiple-Choice Questions (page 26)

1. **D** 24
2. **C** 1, 3, and 9
3. **A** 10 and 25
4. **D** 8
5. **D** 72
6. **C** 8

Adding Fractions

Try This (page 28)

1. $\dfrac{13}{16}$

2. $1\dfrac{8}{35}$

3. $1\dfrac{1}{18}$

4. $\dfrac{17}{20}$

5. $1\dfrac{7}{40}$

6. $1\dfrac{17}{48}$

Subtracting Fractions

Try This (page 29)

1. $\dfrac{7}{16}$

2. $\dfrac{13}{35}$

3. $\dfrac{13}{18}$

4. $\dfrac{19}{60}$

5. $\dfrac{23}{40}$

6. $\dfrac{1}{48}$

Addition and Subtraction with Mixed Fractions

Try This (page 32)

1. $6\dfrac{13}{16}$

2. $3\dfrac{13}{35}$

3. $11\dfrac{1}{18}$

4. $2\dfrac{19}{60}$

5. $7\dfrac{7}{40}$

6. $1\dfrac{1}{48}$

Multiplying Fractions

Try This (page 33)

1. $\dfrac{15}{128}$

2. $\dfrac{12}{35}$

3. $\dfrac{4}{27}$

4. $\dfrac{7}{45}$

5. $\dfrac{21}{80}$

6. $\dfrac{11}{24}$

Dividing Fractions

Try This (page 34)

1. $3\dfrac{1}{3}$

2. $\dfrac{15}{28}$

3. $\dfrac{3}{16}$

4. $2\dfrac{3}{16}$

5. $\dfrac{12}{35}$

6. $\dfrac{32}{33}$

Multiplication and Division with Mixed Fractions

Try This (page 36)

1. $10\dfrac{127}{128}$

2. $1\dfrac{59}{60}$

3. $27\dfrac{47}{54}$

4. $2\dfrac{3}{136}$

5. $7\dfrac{51}{80}$

6. $1\dfrac{49}{176}$

LESSON 3—DECIMAL PLACE VALUE AND COMPUTATION

Think About This (page 37)

1. 3 millionths or 0.000003

2. 5 hundred or 500

3. 7 thousandths or 0.007

4. and

5. decimal point

Multiple-Choice Questions (page 38)

1. D 5 thousandths

2. D ten-thousandths column

3. C two thousand five ten thousandths

4. A 0

Open-Ended Questions (page 39)

1. Whole numbers do not need decimal points.

2. We would use a decimal point if it were money, and we would place it after the 4 and before any cents.

Adding Decimals

Try This (page 39)

1. 790.418

2. 6,520

3. 36.18187

4. 8,579.769

5. 2.1228

6. 10.34567

Subtracting Decimals

Try This (page 40)

1. 722.178

2. 6,424.26

3. 177.67507

4. 8,110.249

5. 3.8968

6. 9.65433

Multiplying Decimals

Try This (page 42)

1. 1,920.6148

2. 567.427

3. 29.666

4. 3,757.6

5. 0.0576

6. 12,750.6

7. 198.2772

8. 12.62058

9. 4.14804

Dividing Decimals

Try This (page 43)

1. 1.6

2. 51.5

3. 282.4

4. 4.8

5. 576

6. 2.1

7. 333.8

8. 28.6

9. 4,000

LESSON 4—PERCENTS

Think About This (page 45)

Percent	Decimal	Fraction as Hundredths	Fraction in Simplest Form
50%	0.5	$\dfrac{50}{100}$	$\dfrac{1}{2}$
15%	0.15	$\dfrac{15}{100}$	$\dfrac{3}{20}$
56%	0.56	$\dfrac{56}{100}$	$\dfrac{14}{25}$
28%	0.28	$\dfrac{28}{100}$	$\dfrac{7}{25}$
85%	0.85	$\dfrac{85}{100}$	$\dfrac{17}{20}$
75%	0.75	$\dfrac{75}{100}$	$\dfrac{3}{4}$
25%	0.25	$\dfrac{25}{100}$	$\dfrac{1}{4}$
40%	0.4	$\dfrac{40}{100}$	$\dfrac{2}{5}$
100%	1.0	$\dfrac{100}{100}$	1

LESSON 5—INTEGERS

Multiple-Choice Questions (page 46)

1. C 36

2. B 0

3. D 25

4. B >

Extended Constructed-Response Question (page 47)

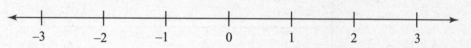

Short Constructed-Response Question (page 47)

−6

Absolute Value

Try This (page 48)

1. 15

2. 8

3. 27

4. 13

5. 4

6. 0

7. −2,800

8. 2,800

LESSON 6—RATIONAL NUMBERS

Try This (page 49)

1. 0.75

2. $-\dfrac{3}{4}$

3. 1.5

4. −1.25

5. $1\dfrac{3}{4}$

6. 0

Try This (page 50)

1. <

2. >

3. <

4. >

5. >

6. <

Rational Numbers on the Coordinate Graph

Try This (page 51)

1.–4. See the following graph.

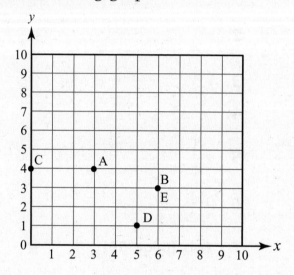

Four-Quadrant Graph

Try This (page 52)

1.–6. See the following graph.

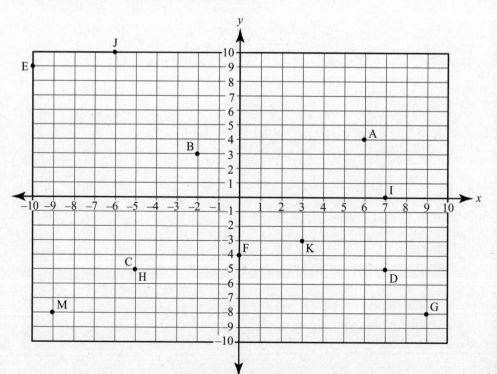

7. (3, –3)

8. (0, 6)

9. (–9, –8)

10. (–7, 0)

11. (–8, 7)

12. (9, 8)

CHAPTER 2—GEOMETRY

LESSON 1—GEOMETRY REVIEW

(page 55)

1. **B** point A; a point is a location in space.

2. **B** pentagon; a pentagon is a five-sided figure.

3. **D** vertical

4. **A** octagon; an octagon has eight sides.

5. **B** diameter

6. **D** segment

7. **A** congruent

8. **C** similar

9. **A** A cube has six faces.

Lines

Try This (page 59)

1. intersecting

2. parallel

3. perpendicular

4. perpendicular

5. intersecting

6. parallel

Triangles

Multiple-Choice Questions (page 61)

1. **A** they have three sides and three angles

2. **C** 180 degrees

3. **B** 60 degrees; $80 + 40 + 60 = 180$

4. **B** 90 degrees; $45 + 45 = 90$; $90 + 90 = 180$

Open-Ended Question (page 62)

Tommy is right. The sum of the angles must equal 180 degrees. $60 + 70 + 80 = 210$ degrees. It is impossible to make a triangle like this.

Classifying Polygons

Think About This (page 62)

1. triangle = 3 sides

2. quadrilateral = 4 sides

3. pentagon = 5 sides

4. hexagon = 6 sides

5. heptagon = 7 sides

6. octagon = 8 sides

7. nonagon = 9 sides

8. decagon = 10 sides

9. A square is a rectangle and a regular polygon.

Think About This (page 63)

1. Answers will vary.

2. cube—6

3. cylinder—3

4. rectangular prism—6

5. Answers will vary.

LESSON 2—UNITS OF MEASUREMENT

Multiple-Choice Questions (page 65)

1. **B** 1 foot 9 inches

2. **C** 800 centimeters

3. **D** about 1 gallon

4. **A** 4 months, 2 weeks, and 5 days

5. **C** about 4 miles

6. **A** 8.5 kilometers

7. **A** 5 degrees

8. **B** 12 ounces

9. **B** 235 grams

10. **B** 4

Open-Ended Question (page 67)

A correct response should include sketches of a cube and a rectangular prism, labeled correctly, along with the response that the cube would hold more because its volume is 216 cubic inches as opposed to the 200 cubic-inch rectangular box.

LESSON 3—MEASURING GEOMETRIC OBJECTS

Circumference

Think About This (page 70)

You would use $\dfrac{22}{7}$ when the diameter is a multiple of 7.

Multiple-Choice Questions (page 70)

1. D 28.8 centimeters

2. B 23.5 inches

3. B 16 feet

4. C 21 yards

5. C 16 inches

Extended Constructed-Response Question (page 71)

They need $30 + 30 + 30 + 30 = 120$ feet of fencing for the yard and $16 \times 3.14 = 50.24$ feet of fence for the pool. The total needed is $120 + 50.24 = 170.24$ feet of fence.

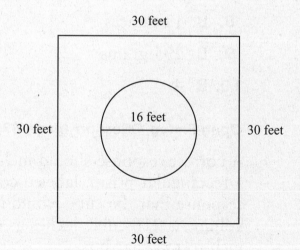

Area of Squares and Rectangles

Sketches may vary.

Try This (page 72)

1. 1,768 cm²

2. 2,064.35 ft²

3. 69.375 mm² or $69\frac{3}{8}$ mm²

4. 1,562.5 mi² or $1,562\frac{1}{2}$ mi²

5. 0.7429 cm²

6. 12 mm²

Try This (page 73)

1. 24 m

2. 300 m

3. 32 m

4. 3.75 m

5. 100 m

6. 450 m

Area of a Parallelogram

Sketches may vary.

Try This (page 74)

1. 1,075 cm²

2. 4,479.14 ft²

3. 103.125 mm²

4. 4,622.5 mi²

5. 0.7136 cm²

6. 14 mm²

Try This (page 75)

1. 20 m

2. 150 m

3. 192 m

4. 2.5 m

5. 125 m

6. 250 m

Area of a Triangle

Try This (page 76)

1. Base: 138 m, Height: 16 m, Area: 1,104 m^2

2. Base: 19.7 cm, Height: 27.4 cm, Area: 269.89 cm^2

3. Base: $15\frac{5}{8}$ in, Height: $10\frac{2}{5}$ in, $81\frac{1}{4}$ in^2

4. 4,872 cm^2

5. 11.934 cm^2

6. $9\frac{1}{6}$ cm^2

Try This (page 78)

1. Base: 128 mm, Height: 36 mm, Area: 2,304 mm^2

2. Base: 77.9 in, Height: 12.3 in, Area: 479.085 in^2

3. Base: $18\frac{1}{2}$ m, Height: $11\frac{1}{4}$ m, Area: $104\frac{1}{16}$ m^2

4. 19,337.5 cm^2

5. 102.785 cm^2

6. $19\frac{1}{4} \text{ cm}^2$

Area of Special Quadrilaterals and Polygons

Try This (page 81)

1. Area of rectangle: 192 cm^2, Area of first triangle: 40 cm^2, Area of second triangle: 40 cm^2, Area of trapezoid: 272 cm^2

2. Area of rectangle: 108 m^2, Area of first triangle: 13.5 m^2, Area of second triangle: 13.5 m^2, Area of trapezoid: 135 m^2

3. Area of rectangle: 754.4 m^2, Area of first triangle: 137.35 m^2, Area of second triangle: 137.35 m^2, Area of trapezoid: $1{,}029.1 \text{ m}^2$

Try This (page 82)

1. 272 cm^2

2. 135 m^2

Areas of Irregular Polygons (page 83)

1. $A = 57 \text{ m}^2$

2. $A = 124 \text{ in}^2$

3. $A = 375 \text{ cm}^2$

Area of a Circle

Multiple-Choice Questions (page 85)

1. **D** 519.84 square centimeters

2. **A** 4 inches

3. **C** 128.7 square centimeters

4. **D** parallelogram

5. **B** $1\frac{11}{16}$ square inches

6. **C** 314 square centimeters

7. **C** 12.56 square kilometers

8. **D** 14 square feet

Open-Ended Question (page 87)

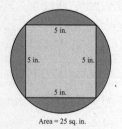

5 in.

5 in.   5 in.

5 in.

Area = 25 sq. in.

The drawing should contain a circle with an inscribed rectangle or square. The length and width of the photograph should measure about $5\frac{1}{2}$ inches or less, and the area should be calculated correctly, based on the dimensions.

Surface Area (page 88)

1.

5 in.

4 in.

3 in.

Think About This (page 89)

2. top = 90 square feet

3. bottom = 90 square feet

4. front = 50 square feet

5. back = 50 square feet

6. left = 45 square feet

7. right = 45 square feet

8. total surface area = 370 square feet

Using Nets to Solve for Surface Area

Try This (page 90)

1.   Diagrams may vary.

1.5 in.

Area of one face: 2.25 in^2, Surface area of cube: 13.5 in^2

2.   Diagrams may vary.

35 mm

30 mm

25 mm

Area of one face: 750 mm^2, Area of second face: 1,050 mm^2, Area of third face: 875 mm^2. Double these amounts. Surface area of rectangular prism: 5,350 mm^2

Surface Area of a Cylinder

Think About This (page 92)

1. A cylinder has three surfaces.

2. A cylinder has two circles and one rectangle.

3. A = 3.14 × 5 × 5; Area = 78.5 square feet

4. circumference

5. 8 feet

6. 31.4 feet

7. 31.4 feet

8. 31.4 × 8 = 251.2 square feet

9. 78.5 + 78.5 + 251.2 = 408.2 square feet

10. square feet

Multiple-Choice Question (page 93)

1. C radius

2. B bottom

3. C 286 square meters

4. A circle

5. D rectangle

6. C circumference

7. D 175.84 square feet

Open-Ended Question (page 95)

The cube requires more paper. It has a surface area of 384 square inches (8 inches × 8 inches × 6 sides) and the rectangular prism has a surface area of 376 square inches (48 + 48 + 60 + 60 + 80 + 80).

Volume

Try This (page 96)

1. $7 \times 6 \times 8 = 336$

2. cubic inches

Try This (page 97)

1. B: 672 mm^2, V: 91,392 mm^3

2. B: 566.1 in^2, V: 4,981.68 in^3

3. B: $16\dfrac{3}{16}$ cm^2, V: $161\dfrac{7}{8}$ cm^3

Volume of a Cylinder

Think About This (page 99)

1. circle

2. $A = \pi r^2$

3. volume equals area of the base times height

4. 153.86 square inches or 154 in^2 when $\pi = \dfrac{22}{7}$

5. 461.58 cubic inches or 462 cubic inches when $\pi = \dfrac{22}{7}$

6. cubed units

Multiple-Choice Questions (page 99)

1. B 35 cubic feet

2. A 8 meters

3. D 452.16 square centimeters

4. C 1 cubic foot

5. B 9.42 cubic meters

Open-Ended Question (page 100)

The answer should indicate that the volume of the rectangular container is equal to 120 cubic feet, and the volume of the cube is equal to 125 cubic feet. Therefore, Matt should buy the cube.

Polygons on the Coordinate Graph

Think About This (page 101)

1. B. (6,3)

   C. (0,4)

   D. (5,1)

2.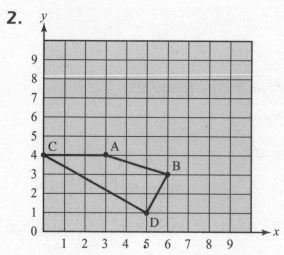

3. Trapezoid

4.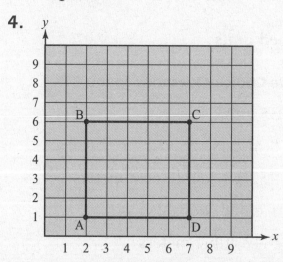

   A. (2,1)

   B. (2,6)

   C. (7,6)

   D. (7,1)

   Answers will vary.

5.

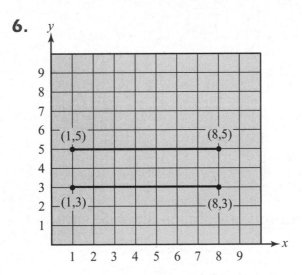

Answers will vary.

6. Answers will vary.

CHAPTER 3—EXPRESSIONS AND EQUATIONS

LESSON 1—PATTERNS, MODELS, AND GRAPHS

Multiple-Choice Questions (page 104)

1. **D** 23

2. **B** 3

3. **C** 5.65

4. **A** −4

Think About This (page 106)

1. $n + 5 = 9$ when $n = 4$

2. $n - 0.8 = 5.8$ when $n = 6.6$

Try This (page 106)

1.

n	$n \times 5$
0	0
3	15
7	35
10	50
14	70

2.

n	$1\frac{1}{8}$	$3\frac{3}{8}$	$5\frac{1}{8}$	7	9
$n - \frac{1}{8}$	1	$3\frac{1}{4}$	5	$6\frac{7}{8}$	$8\frac{7}{8}$

3. Multiply a number times 5; subtract $\frac{1}{8}$ from n.

4. $5n$ or $n \times 5$; $n - \frac{1}{8}$.

Tables

Try This (page 108)

1. $n + 7$

2. $n - 2$

3. $n \times -1$

4. $2n + 1$

Models

Multiple-Choice Questions (page 109)

1. C 12

2. A $2 + n = 9$

3. C Four times the sum of three plus n is equal to 16

4. C 6

5. B 4

Graphs

(page 113)

1.

x	$3x + 1$
0	1
1	4
2	7
3	10

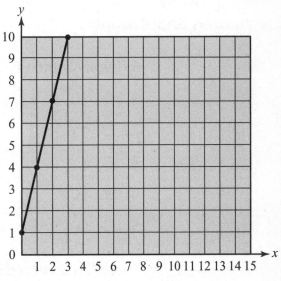

LESSON 2—THE DISTRIBUTIVE PROPERTY

Fill in the Blanks (page 114)

$7 \times (6 + 3) =$

$7 \times 9 = 63$

$(5 \times 7) - (5 \times 2) =$

$5 \times (7 - 2) =$

$5 \times 5 = 25$

Multiple-Choice Questions (page 115)

1. B 6×6

2. A 8×2

3. D 81

4. B 42

Short Constructed-Response Questions (page 116)

6. $21 + 35 = 7(3 + 5)$
   $56 = 7 \times 8$

7. $6x$

LESSON 3—EVALUATING EXPRESSIONS AND INEQUALITIES

Evaluating Expressions

Multiple-Choice Questions (page 117)

1. C 20

2. D 45

3. A 4

4. B 100

5. D 40

Short Constructed-Response Questions (page 118)

6. 343

7. 13

Inequalities

Multiple-Choice Questions (page 119)

1. B $6 + 3 \leq 10$

2. A 4

3. B $2 \times n = 12$

4. C $4 \times n \geq 32$

Short Constructed-Response Questions (page 120)

5.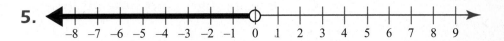

6.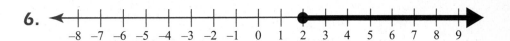

Extended Constructed-Response Question (page 120)

Mr. Johnson—$24.00

Mrs. Johnson—$24.00

Jerome—$18.00

Luisa—$18.00

William—$12.00

Tina—Free

Sammy—Free

Total = $96.00

CHAPTER 4—STATISTICS AND PROBABILITY

LESSON 1—UNDERSTANDING DATA

Bar Graphs

Multiple-Choice Questions (page 123)

1. D 100
2. A 5
3. D 55
4. C apple, mango, and pineapple
5. C mango and nectarine

Open-Ended Question (page 125)

Answers will vary.

Circle Graphs

Multiple-Choice Questions (page 126)

1. C multiply by 2
2. D 44 percent
3. C 108 degrees
4. D 90 percent
5. C 216 degrees

Open-Ended Question (page 128)

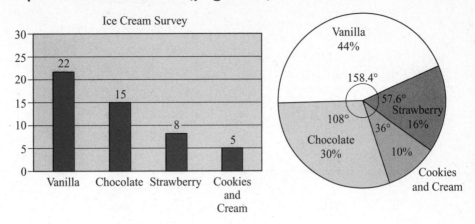

Additional Graphs

Try This (page 131)

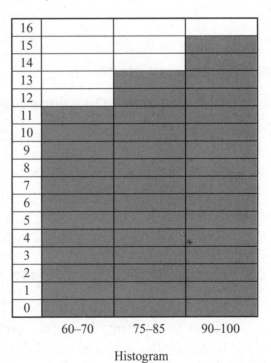

Histogram

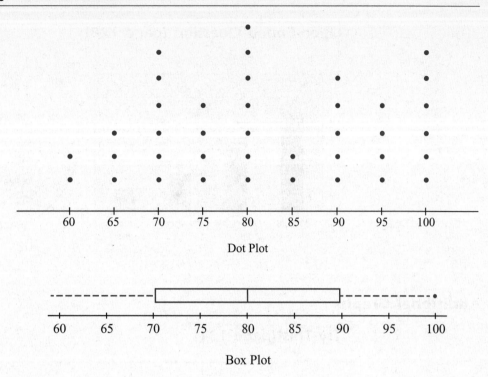

Dot Plot

Box Plot

Range, Mean, Median, and Mode

Multiple-Choice Questions (page 133)

1. **B** 14 degrees

2. **C** 61 degrees

3. **D** 62 degrees

4. **D** 62 degrees

Open-Ended Question (page 134)

The mean changed because the sum of the temperatures and the divisor changed. The range remained the same because the smallest and largest numbers were the same, the median remained the same because 62 was still the middle number in the set, and the mode remained the same because 62 still appeared the greatest number of times in the set.

LESSON 2—PROBABILITY

Multiple-Choice Questions (page 135)

1. B $\frac{1}{8}$

2. A 0

3. D 1

4. C $\frac{1}{2}$

Open-Ended Question (page 137)

Answers will vary.

Experimental Probability

Multiple-Choice Questions (page 138)

1. C $\frac{3}{5}$

2. B $\frac{2}{5}$

3. D $\frac{4}{5}$

4. D $\frac{8}{30}$

5. A $\frac{1}{5}$

CHAPTER 5—RATIOS AND PROPORTIONAL RELATIONSHIPS

LESSON 1—RATIOS AND PROPORTIONS

Think About This (page 141)

1. 12 to 36 or 12:36 or $\dfrac{12}{36}$ or $\dfrac{1}{3}$

2. 12 to 60 or 12:60 or $\dfrac{12}{60}$ or $\dfrac{1}{5}$

3. 3 to 4 or 3:4 or $\dfrac{3}{4}$

Think About This (page 142)

1. 26 black cards

2. 8 candy bars

3. 120 minutes

4. 2 tacos

Multiple-Choice Questions (page 143)

1. B 5:26

2. D 24:7

3. C 1:1

4. C 8

5. C 20

6. C 18 inches

7. C $4.50

Open-Ended Question (page 145)

No. The enlarged picture would be 12 × 20 inches and the frame measures 12 × 18 inches.

LESSON 2—UNIT RATES

Try This (page 146)

1. $0.49

2. $25

NEW JERSEY ASSESSMENT OF SKILLS AND KNOWLEDGE 2007 GRADE 6 MATHEMATICS REFERENCE SHEET

Use the information below to answer questions on the Mathematics section of the 2007 Grade Six Assessment of Skills and Knowledge (NJ ASK 6).

The sum of the measures of the interior angles of a triangle = 180°

Distance = rate × time

Simple Interest Formula: $A = p + prt$

A = amount after t years; p = principal; r = annual interest rate; t = number of years

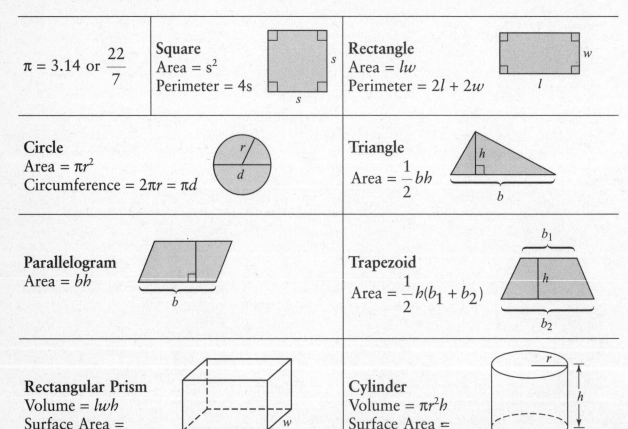

$\pi = 3.14$ or $\dfrac{22}{7}$

Square
Area = s^2
Perimeter = 4s

Rectangle
Area = lw
Perimeter = $2l + 2w$

Circle
Area = πr^2
Circumference = $2\pi r = \pi d$

Triangle
Area = $\dfrac{1}{2}bh$

Parallelogram
Area = bh

Trapezoid
Area = $\dfrac{1}{2}h(b_1 + b_2)$

Rectangular Prism
Volume = lwh
Surface Area =
    $2lw + 2wh + 2lh$

Cylinder
Volume = $\pi r^2 h$
Surface Area =
    $2\pi rh + 2\pi r^2$

USE THE FOLLOWING EQUIVALENTS FOR YOUR CALCULATIONS

60 seconds = 1 minute	12 inches = 1 foot	10 millimeters = 1 centimeter
60 minutes = 1 hour	3 feet = 1 yard	100 centimeters = 1 meter
24 hours = 1 day	36 inches = 1 yard	10 decimeters = 1 meter
7 days = 1 week	5,280 feet = 1 mile	1,000 meters = 1 kilometer
12 months = 1 year	1,760 yards = 1 mile	
365 days = 1 year		

8 fluid ounces = 1 cup	16 ounces = 1 pound
2 cups = 1 pint	2,000 pounds = 1 ton
2 pints = 1 quart	1,000 milligrams = 1 gram
4 quarts = 1 gallon	100 centigrams = 1 gram
1,000 milliliters (mL) = 1 liter (L)	10 grams = 1 dekagram
	1,000 grams = 1 kilogram

MATHEMATICS MANIPULATIVES SHEET
SHAPES, PROTRACTOR, AND RULER
GRADES 5, 6, AND 7 ONLY

Regular Triangles

Ruler

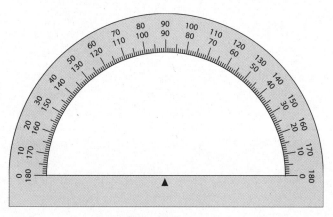

Protractor

INDEX